BIOGEOCHEMICAL CYCLES AND CLIMATE

Biogeochemical Cycles and Climate

Han Dolman

Vrije Universiteit Amsterdam

OXFORD

UNIVERSITY PRESS

Great Clarendon Street, Oxford, OX2 6DP,
United Kingdom

Oxford University Press is a department of the University of Oxford.
It furthers the University's objective of excellence in research, scholarship,
and education by publishing worldwide. Oxford is a registered trade mark of
Oxford University Press in the UK and in certain other countries

First published 2019
First published in paperback 2021

Published in the United States of America by Oxford University Press
198 Madison Avenue, New York, NY 10016, United States of America

British Library Cataloguing in Publication Data
Data available

Library of Congress Cataloging in Publication Data
Data available

ISBN 978–0–19–877930–8 (Hbk.)
ISBN 978–0–19–284526–9 (Pbk.)

Printed and bound by
CPI Group (UK) Ltd, Croydon, CR0 4YY

Preface

It was at the annual meeting of the European Geosciences Union in Vienna in 2014 that I was approached by Sonke Adlung from Oxford University Press with the question if any and, if so, what book was missing in my field of research. I think now that I did not immediately reply, but that it took me some time before I was able to clearly identify what was missing. The reasons for that lie in my experience of having taught Earth science students the basic physics of climate for some time, and in the fact that Earth science itself provides numerous beautiful examples of how that physics and biogeochemistry operated in the geological past, aspects which I did not teach. I needed some time to formulate the importance of presenting these together.

Most university courses present our students with a rather incomplete picture of the climate of our planet. Meteorology students are made familiar with the physics of the interaction of greenhouse gases and radiation, and the thermodynamics and transport of the atmosphere, but generally lack knowledge of the role of oceans and the interaction of the biosphere with climate. They also miss a palaeoclimatological perspective. Earth science students gain insight into long-term climate processes such as the geological thermostat, the role of plate tectonics in climate and various other aspects of palaeoclimate and biogeochemistry but are hardly made familiar with the physics of atmosphere and ocean.

The societal implications of this development are severe. We can only understand climate (change) as long as we can quantify the multitude of feedbacks between these important physical and biogeochemical and biological processes. This requires understanding not only what processes change the rates and magnitudes of biogeochemical cycles such as the carbon cycle, the nitrogen cycle and the water cycle but also how the physics of motion, thermodynamics and radiation respond to those changes.

This book follows from these considerations. It is about the science of the interactions and the exchange processes between reservoirs such as the ocean, the atmosphere, the geosphere and the biosphere. It is less about the human impact, as there is already a wealth of good textbooks that deal with issues of current climate change such as adaptation, mitigation and impacts. The book is primarily aimed at Earth science students who are at the advanced-bachelor's or master's level and have a basic understanding of algebra, chemistry and physics. Its purpose is not at all to be complete. It aims rather to provide a more integrated view of the climate system from an Earth science perspective. The choices of the subjects are, however, my own and based on personal and probably biased preferences. It is up to the reader to judge how well I succeeded in this complicated attempt at integration.

The first three chapters offer a general introduction to the context of the book, outlining the climate system as a complex interplay between biogeochemistry and physics and describing the tools we have to understand climate: observations and models. They describe the basics of the system and the biogeochemical cycles. The second part consists of four chapters that describe the necessary physics of climate to understand the interaction of climate with biogeochemistry and change. The third part of the book deals with Earth's (bio)geochemical cycles. These chapters treat the main reservoirs of Earth's biogeochemical cycles—atmosphere, land and ocean—together with their role in the cycles of carbon, oxygen, nitrogen, iron, phosphorus, oxygen, sulphur and water and their interactions with climate. The final two chapters describe possible mitigation and adaptation actions, always with an emphasis on the biogeochemical aspects. I have tried to make these last six chapters as up to date as possible by providing more references in them than in the previous chapters.

I am grateful to Sonke Adlung for posing that initial question to me in 2014. He and Ania Wronski at Oxford University Press have always been very supportive of the project, even when the start was somewhat difficult. John Gash, former colleague and lifelong friend, edited the first draft. Various colleagues at the Department of Earth Science of the Vrije Universiteit have provided feedback, both on ideas and on chapters. Gerald Ganssen has always been a great stimulating force for this project and I very much appreciate his role as discussion partner providing the larger-scale geological perspective. I sincerely acknowledge the freedom I experienced in the Department of Earth Sciences to be able to write and finish this book as I wished. Further support was received from the Darwin Centre for Biogeology and the Netherlands Earth System Science Centre. Several people have commented on individual chapters: Antoon Meesters, Jan Willem Erisman, Joshua Dean, Sander Houweling Nick Schutgens, Appy Sluys and Jack Middelburg. Ingeborg Levin and Martin Heimann read the near-final draft. It is fully due to these people that several (embarrassing) errors have been corrected in time. I very much appreciate the opportunity provided by the president of the Nanjing University of Information Science and Technology and by Tiexi Chen and Guojie Wang to spend some time at Nanjing in the spring of 2018, that allowed me to finish the book. Kim Helmer was of great help in obtaining copyright permissions for the various figures used in this book.

The writing of this book took quite some time, but was overall a very enjoyable experience. I like to thank my lifelong partner Agnes and my two sons Jim and Wouter for their support. Words are simply not enough to express my gratitude to them.

Han Dolman
June 2018

Contents

1

Introduction

1.1 Biogeochemical cycles, rates and magnitudes

The movement of matter and transfer of energy around the planet plays a fundamental role in Earth's system. It ensures that Earth is habitable and also determines the availability of key resources for human use. Take, for instance, the availability of fossil fuels such as the hydrocarbons oil and gas. The large-scale exploitation of these fossil fuel reserves has enabled Earth's population to develop on an incredibly fast economic growth trajectory that has brought us great wealth and progress. However, the waste products of this fossil fuel use, primarily in the form of carbon dioxide and methane, are causing large-scale perturbations in Earth's climate. These have been assessed by the Intergovernmental Panel for Climate Change (IPCC) in a series of groundbreaking assessment reports, documenting rising sea levels, increasing temperatures, changing weather patterns, heatwaves and drought. In geological history, some of the greenhouse gases that are now playing havoc with the climate have played an essential role in regulating Earth's temperature and created the conditions for life to evolve. At these long timescales, without human interference, the carbon atoms in carbon dioxide and methane molecules would be continuously recycled in Earth's geological cycles. Variations in the rate of biogeochemical cycling are key to understanding past climate variability and today's climate change.

The rate and magnitude of biogeochemical cycling also determine the availability of food for humans, as the large-scale application of artificial fertilizer shows. Haber invented the process for the industrial production of ammonia from atmospheric nitrogen in 1908; since then, many more people on Earth have been fed, particularly in the period after the Second World War. The concomitant population explosion has, however, happened at the cost of substantial changes to the biogeochemical cycle of nitrogen and to the environment, to the extent that now the nitrogen cycle is dominated by input from anthropogenic processes. Environmental effects include contamination and eutrophication of freshwater bodies, and excessive atmospheric deposition and emissions of a potent greenhouse gas, nitrous oxide.

The Russian scientist Vladimir Vernadsky is generally credited with defining the notion of the biosphere in his landmark 1926 book, the Biosphere and the associated biogeochemical cycling of material and energy within the biosphere (Vernadsky, 2012). In his view, the biosphere is one of Earth's concentric envelopes, containing not only life

Biogeochemical Cycles and Climate. Han Dolman, Oxford University Press (2019). © Han Dolman.
DOI: 10.1093/oso/9780198779308.001.0001

but also the minerals that are cycled through the activity of life and geological processes. He was the first to entertain the notion that the composition of the atmosphere was in some way regulated by life. One could argue that he identified living matter as a geological force in its own right, a concept that has now found its human equivalent in the modern concept of the Anthropocene (Crutzen, 2002).

The impact of biogeochemical cycles on climate and vice versa can conveniently be phrased as three questions:

- How have the cycles of key nutrients, such as carbon, nitrogen and phosphorus, and water changed, both in the geological past and, more recently, through the impact of humans on the Earth system?
- How do these cycles interact with each other and the physical properties of climate?
- How can we use this knowledge to mitigate some of the impacts of the changing biogeochemistry on climate, and Earth's habitability and resilience?

This book is about these three aspects of biogeochemical cycling and Earth's climate. Understanding the exchange of materials and its relation to climate is important, in particular if these exchanges involve radiatively active trace gases such as carbon dioxide, methane and nitrous oxide. Through their absorption characteristics in the infrared radiation domain, these trace gases directly interact with the climate.

While the geological forces that have shaped the environment over millions and billions of years impact the size and availability of fluxes and reservoirs, geology also provides key signature information on changes in biogeochemical cycles, through long-term records of the composition and abundance of specific minerals and isotopes in sediments and rocks. These records are not always straightforward or easy to interpret; however, the geological memory is an important asset in providing the larger picture of the interaction between biogeochemistry and climate. We will take this double-sided perspective in this book: Earth science is both our tool and our study object.

1.2 The geological cycle

One of the important characteristics of Earth is its continuous recycling of rocks, soils and material through what is known as the geological cycle. The geological cycle is traditionally described as a set of four related cycles: the tectonic cycle, the rock cycle, the hydrological cycle and the biogeochemical cycles. The last two figure prominently in this book. Virtually all of these cycles are, in some way, interlinked but they differ in the timescales at which key processes occur. The tectonic cycle involves the creation and destruction of the lithosphere (Earth's outer layer, which is roughly 100 km deep) through a process called plate tectonics. Not all planets have this phenomenon, at present, it appears to be a unique feature of Earth. Planetary geologists still debate why Earth is the only planet in the solar system with active plate tectonics. Plate tectonics operates at timescales of tens of millions of years and involves the movement of large plates on a substrate of denser material at speeds of 0.020–0.150 m yr^{-1} depending on the plate and geological period. On the one hand, the tectonic cycle involves the creation of new materials, or lithosphere, at oceanic ridges, in a process called 'sea-floor spreading'. On the other hand, it involves the destruction of these in 'subduction zones', where an oceanic plate dives, or subducts, due to its larger density, beneath a lighter continental plate. Subduction zones are the places with the most active volcanoes and where seismic activity is strongest. Where tectonic plates of similar density meet, mountain ridges are produced as a result of the collision; the Himalayas and the Rocky Mountains are prime examples of this. Plate tectonics shapes the continents, oceans and mountain ridges of the planet, and has an important impact on climate—not least because the placement of continents affects the ocean circulation. Plate tectonics provides the critical recycling mechanism for materials in Earth's crust.

Trace gases such as carbon dioxide and methane, but also water vapour, define the radiative properties of the atmosphere, keeping the global tropospheric temperature well above freezing. On Earth, water can exist in its frozen, liquid and vapour forms and this is a prerequisite for life as we know it. Carbon dioxide is only present in Earth's atmosphere in small quantities, with its concentration typically expressed in parts per million. This is fundamentally different to our neighbouring planets Venus and Mars, which both have atmospheres comprising more than 95% carbon dioxide. It is estimated that, if there had not been life on Earth, its atmosphere would contain 98% carbon dioxide. The amount of carbon in Earth's atmosphere is, however, marginal compared to the quantities of carbon locked up in the deep Earth, sediments and ocean reservoirs.

It is worth emphasizing these differences between the carbon stocks of the surface reservoirs and the stores of carbon in Earth's core, mantle and continents. Earth's core is estimated to contain 4 billion petagrams of carbon (Pg C; 1 Pg = 10^{15} grams), which is 90% of the total amount of Earth's carbon, an enormous 1 million times more than the three surface reservoirs together (DePaolo, 2015). This deep carbon does not exchange with the other reservoirs but has probably played an important role in establishing climate on the very young Earth, some 4.5 billion years ago. Earth's mantle contains about 240–400 million Pg C, roughly 8%–9% of the total amount of Earth's carbon. This carbon appears as diamonds and impurities in minerals. About a quarter, 60–70 million Pg, of this is located in Earth's crust, forming the continents and the ocean floor. Its main forms are limestone and dolomite, and its pressurized form is marble. This carbon may be released from the crust to the atmosphere through volcanism and geysers.

1.3   The carbon cycle

To illustrate the impact of biogeochemistry on climate, let us pay some further attention to the carbon cycle. The biggest game-changing event in Earth's biogeochemical history was arguably the production of organic matter and oxygen by cyanobacteria, algae and green plants (Langmuir & Broecker, 2012). In the presence of solar radiation,

$$6CO_2 + 6H_2O \rightarrow C_6H_{12}O_6 + 6O_2 \tag{1.1}$$

This equation is fundamental to our understanding of the carbon and oxygen cycle, past and present, and we will come back to it numerous times in this book. It describes the production of oxygen and sugars through oxygenic photosynthesis. The subsequent reverse process, respiration, consumes oxygen and sugars in an exothermic reaction. To a large extent, the balance of these two processes determines the size of surface reservoirs of carbon and oxygen. However, interactions with other geological processes, such as reactions with reduced materials from Earth's crust (e.g. sulphur and iron) or the burial of organic matter, can take carbon temporarily out of the equation and shift the equilibrium towards the right, thus increasing the amount of oxygen. Of course, the opposite can also happen. That part of the carbon cycle where this equation plays the central role is named the organic carbon cycle (see Figure 1.1). We will see later that its importance in regulating climate depends on the timescale we look at and that it cannot explain all the variability of the geological record (see Chapter 9). For that, we also need to look at another part of the carbon cycle.

This other part, the inorganic carbon cycle, is often called Earth's geological thermostat and is shown schematically in Figure 1.2. It describes how, in its solid form, calcium carbonate is in a subtle exchange process with carbon dioxide in the atmosphere. The rate of exchange between atmosphere, ocean and crust appears to control climate at million-year timescales. It is dependent—as the name suggests—on the geological cycle, but it is also dependent on the availability of liquid water in the hydrological cycle. Weathering of exposed rock surfaces, either physically, by freezing and cracking, or chemically, by dissolution, produces smaller particles of these rocks; these particles are then transported by wind, water or ice. When these materials collate together and sink to the floor of the ocean basins, sediments are produced. The weight of the new sediments

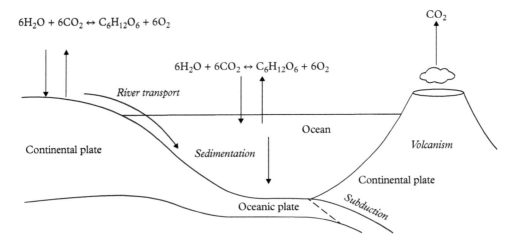

Figure 1.1 *The organic carbon cycle based on the conversion of carbon dioxide and water into oxygen and sugars (biomass). The speed at which the sediments are buried and subsequently returned to the atmosphere through subducting plates that emit volcanic carbon dioxide links the organic carbon cycle to the tectonic cycle.*

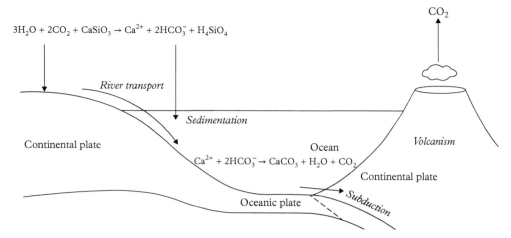

Figure 1.2 *The inorganic carbon cycle containing the geological thermostat. The speed at which the sediments are buried and subsequently returned to the atmosphere through subducting plates that emit volcanic carbon dioxide then determines the net long-term balance of carbon dioxide in the atmosphere.*

compacts the underlying layers further until sedimentary rocks are produced. These then enter the tectonic cycle and may be transformed by heat and pressure into metamorphic rocks or lifted up to the surface by plate tectonics, where a new cycle of weathering can start. The ocean, in particular the deep ocean, thus plays a key role in linking the fast and slow carbon cycles.

A closer look at the chemistry of water, carbon and the rocks of the inorganic carbon cycle shows the fundamental importance of this part of the carbon cycle on geological timescales (Langmuir & Broecker, 2012). Water falling on Earth's land surface lets rocks and soil react chemically with the main minerals, calcium and silicate:

$$3H_2O + 2CO_2 + CaSiO_3 \rightarrow Ca^{2+} + 2HCO_3^- + H_4SiO_4 \tag{1.2a}$$

$$Ca^{2+} + 2HCO_3^- \rightarrow CaCO_3 + H_2O + CO_2 \tag{1.2b}$$

Equation (1.2a) shows how water and carbon dioxide act together to dissolve the calcium silicate (the mineral wollastonite) to produce calcium and carbonate ions and a silicate. In today's ocean, calcium is used to form the shells of small unicellular and multicellular organisms that ultimately sink down to the ocean floor. In the absence of these organisms, the calcium and bicarbonate ions react to give calcium carbonate that then precipitates to the ocean floor (eqn (1.2b)).

The reason why these equations are so important in controlling Earth's climate is that the dissolution of one molecule of wollastonite occurs at the cost of precisely one molecule of carbon dioxide (note that there are two molecules of carbon dioxide going in on the left side of eqn (1.2a) but only one coming out on the right side of eqn (1.2b)). This provides a mechanism for the effective removal of carbon dioxide from the atmosphere by Earth's crust. But this is not the full story: eqn (1a) also shows the importance of water. If there is more water, that is, in the form of precipitation, the weathering rate should go up and thus also the amount of carbon dioxide removed from the atmosphere. Higher temperatures can lead to increased precipitation by increasing the water-holding capacity of the atmosphere. If the atmospheric concentration of carbon dioxide increases, the temperature goes up and thus more carbon dioxide is removed. If the temperature goes down, the opposite happens, and the carbon dioxide concentration in the atmosphere, and the temperature, will go up. This feedback loop is thought to be the main regulator of Earth's climate at geological timescales and shows how intricately water, the carbon cycle and climate are linked: the main thesis of this book. The fact that it operates at a geological timescale makes testing and finding hard evidence problematic, so it is best considered a (very) likely hypothesis (but see also Chapter 12).

The inorganic carbon cycle operates at timescales of millions to billions of years and is of little use to the current increase in carbon dioxide due to the use of fossil fuels. Indeed, on top of this slow inorganic cycle operates a much faster carbon cycle (Ciais et al., 2013). Here, fluxes of carbon involve the uptake of carbon by plants through photosynthesis, its subsequent respiration (eqn (1.1)) both on land and in the ocean, and the uptake of carbon dioxide by the ocean. The fluxes of this part of carbon cycle impact the surface stores of Earth, the carbon reservoirs of ocean, land and atmosphere. By far the largest amount of carbon is stored in the ocean, about 40,000 Pg C. The amount of carbon stored on land is tiny in comparison, a mere 500 Pg C in the biomass and 3–4 times that in the soil (pre-industrial numbers), with an additional 1,700 Pg C locked up in the soils of the permafrost areas, mostly in Siberia (see Chapter 10). In the atmosphere, the third exchange reservoir, the amount stored is comparable to that of biomass on land, about 500 Pg C.

Compare these minute amounts with those we mentioned earlier for the deep Earth and crust; they comprise a mere 1%–2% of the total carbon reserves of Earth. However, the carbon in the smallest of these, the atmospheric reservoir, and in concentrations of carbon dioxide in parts per million, provides the key to climate and how it changes in the short term. In the longer term, the oceans' uptake capacity and their ability to transport carbon dioxide to their deeper layers, the ocean sediments, and, on a much longer scale, tectonic movements and volcanism are the important players.

1.4   Feedbacks and steady states

Feedbacks such as that of the geological thermostat, as well as from the burial of organic matter, play a crucial role in the complex system of Earth's climate. A feedback either dampens or strengthens an original signal. Negative feedbacks stabilize a system; positive feedbacks amplify initial perturbations. For a feedback to occur, several components are needed: an initial signal, a process that responds to this signal, and an amplifying or dampening mechanism. A classic example in climate science is the ice–albedo feedback. Ice has a high albedo and thus reflects a large amount of sunlight, substantially more so than open water or land. The ice surface therefore stays cool relative to open water or land.

If the temperature increases (the signal), more ice can melt (the process in the feedback loop). This results in more open water and ice-free land, and even higher temperatures, which cause even more ice to melt. This kind of feedback, where an initial signal is amplified over and over again, is a positive feedback loop. One example of a negative feedback is growing-season phenology: here, warmer temperatures extend the growing season, leading to more photosynthesis. The increased carbon drawdown can act to reduce the amount of warming. Here, the signal is the temperature of the atmosphere, the process is photosynthesis and the result is a dampening of the initial signal.

In a steady state, the size of the reservoir and the magnitude of the fluxes do not change. Cycles involving exchange fluxes between reservoirs are often depicted with a box model (Figure 1.3). The simplest cycle would involve two reservoirs, S_1 and S_2, between which fluxes are exchanged as $F_{1,2}$ and, vice versa, $F_{2,1}$. In steady state conditions, the following equation holds:

$$\frac{dS_1}{dt} = F_{1,2} - F_{2,1} = 0 \tag{1.3}$$

In practice, all (bio)geochemical cycles have components, or storage reservoirs, on land, in the ocean and in the atmosphere, thus adding a third reservoir to our simple box model. This adds complexity in the form of new exchange and return fluxes, as now six fluxes are important rather than two. It is difficult to generalize this, as adding new components generally involves exchanges with one or two of the other reservoirs only. The interplay between such a set of exchange fluxes determines the complexity and stability of a cycle, as feedbacks may be strengthened or weakened by changes in the size of the fluxes.

An important descriptor of the functioning of a cycle is the ratio between the size of the reservoir (i.e. the mass) and the flux. If the fluxes are small compared to the size of the reservoir, the process generally has little implication at short timescales. The reverse is true if the flux is large compared to the size of the reservoir. This ratio is called the turnover time. This is equivalent to the mean residence time under steady state conditions and represents the average time it takes for a molecule to pass through the reservoir (e.g. Bolin & Rodhe, 1973). The mean residence time, τ, is calculated as the ratio of the mass of the reservoir to the sum of either input or output fluxes and assumes steady state conditions:

$$\tau = \frac{S_1}{F_{1,2}} \tag{1.4}$$

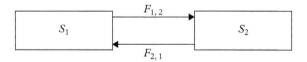

Figure 1.3 *Schematic drawing of a very simple geological cycle with two reservoirs and two reciprocal exchange fluxes*

Table 1.1 *The fluxes, reservoir sizes and calculated the mean residence time (τ) for key reservoirs of the carbon cycle. (Values from Ciais et al., 2013)*

Reservoir	Flux (Pg C yr⁻¹)	Size of Reservoir (Pg C)	τ (years)
Atmosphere	190	589	3.1
Ocean	80	40000	500.0
Land vegetation	123	525	4.3
Soils	47	950	41.0

In Table 1.1 the magnitude of the fluxes and reservoirs of the carbon cycle are given and used to calculate the mean residence times, to indicate the stability of carbon in the reservoirs. As the sizes of the fluxes and the reservoirs are similar, the lowest mean residence times for carbon are obtained for the atmosphere and the vegetation. The ocean reservoir—given its enormous size—has a mean residence time for carbon of 500 years for a molecule of carbon dioxide. It is important to realize that, although the lifetime of an individual molecule of carbon dioxide in the atmosphere is relatively short, in most cases this simply implies that a molecule is taken up by the ocean and land, and another molecule swapped back. For the warming potential of carbon dioxide, the mean residence times of carbon in the reservoirs are thus less important than the absolute amounts.

The processes ensuring the uptake of carbon by vegetation and ocean are critically important in determining the rate of change of the atmospheric carbon stock, and hence climate change. We can illustrate this further by showing the impact of mean residence times on the change of the size of the reservoirs after an initial pulse of carbon dioxide for the normalized change in atmospheric content since the industrial revolution. Figure 1.4 shows how the initial uptake takes place within the reservoir with the fastest response time, in this case, the land system (taken as the mean residence time for soils, from Table 1.1). We assume that

$$\frac{dS}{dt} = \beta S \tag{1.5a}$$

$$S = S_0 e^{-\beta t} \tag{1.5b}$$

which shows that the rate of change in reservoir S, the atmosphere, in our case, is assumed to be proportional to a constant β times the size of the reservoir. The mean residence time is then the reciprocal of this constant. Note also that the mean residence time is equivalent to the time it takes to drop to $0.31S$, that is, $1/e$ of the initial value S_0. In this extremely simplified model, it would thus take the land about two hundred years to get rid of the assumed pulse of anthropogenic carbon dioxide, provided it continues to take up carbon. This is, of course, not quite realistic, because of the ultimately limited capacity of the biosphere to absorb carbon. So, the decline will be much less rapid and, importantly, feedbacks exist between the organic part of the carbon cycle and climate, through the temperature and moisture sensitivity of respiration and photosynthesis,

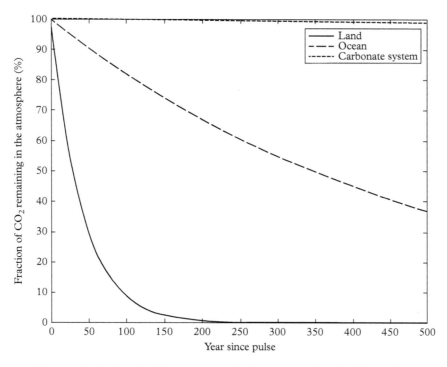

Figure 1.4 *Removal of carbon dioxide, expressed as the fraction of carbon dioxide remaining in the reservoir after an initial pulse (100%) is fed through the main reservoirs: land, deep ocean and the carbonate system. The mean residence times used are based on the values given in Table 1.1 and a simple first-order differential equation in which the removal rate is set proportional to the size of the reservoir and the reciprocal of the mean residence time, that is, eqns (1.5a) and (1.5b).*

which renders this rough calculation even less accurate. The deep ocean removes carbon at much longer timescales. At even longer, geological timescales, inorganic carbonate sedimentation moves carbon from the ocean into the crust: these processes would take several tens or hundreds of thousands of years to remove the pulse of human carbon dioxide. Over the first 500 years after the pulse, this system would only remove 1% of the initial pulse.

This model does not take into account the full buffering capacity of the ocean's carbonate system. This may make the uptake of carbon considerable less than estimated here (see Chapter 9). Despite these caveats, the calculations make several important points. In the long term, Earth will recover from the pulse of fossil fuel emission humans have injected into the atmosphere, initially through the action of both the land and the oceans, and later primarily through the action of the larger reservoir, the ocean. However, limits exist to the uptake capacity and rate of these systems; for the oceans, this will be a very slow process. Interestingly, the uptake of carbon by land and ocean currently takes 'care' of about half our emissions, preventing a much stronger heating of the planet. We will explore this feedback in more detail in Chapters 9 and 13.

1.5   The greenhouse effect and the availability of water

After showing the importance of reservoirs and fluxes in responding to a change in input, let us go back to the role of greenhouse gases in climate. The physics of the greenhouse effect is dealt with in detail in Chapter 4 so, for the moment, a brief introduction will suffice. Gases, particularly those whose molecules comprise more than two atoms, such as carbon dioxide, nitrous oxide, water vapour and methane, have the ability to absorb radiation at specific wavelengths in the infrared domain of the electromagnetic spectrum, that is, wavelengths just longer than those of visible light. We call these gases 'greenhouse gases'. However, they are also able to radiate this radiation back into all directions. So, if the atmosphere is warmed from below by radiation from Earth, the greenhouse gases in the lower atmosphere will absorb part of this radiation and radiate it back to Earth, effectively warming the lower atmosphere. The details of greenhouse gas radiation absorption and emission are undisputed and were discovered in the middle part of the nineteenth century by Fourier, Tyndall and Arrhenius. For now, it is sufficient for us to know that an increase in greenhouse gas concentrations will lead to an increase in temperature. How much exactly this increase is, and how much this impacts, for instance, the hydrological cycle and other processes on Earth, we can blissfully ignore for the moment.

We will use this greenhouse effect of water vapour and carbon dioxide to explain how water vapour, liquid water and frozen water can coexist on Earth (e.g. Kasting, 2010). The availability of water in these three states is essential for life as we know it: too cold an atmosphere will yield only ice, while too hot an atmosphere will contain only water vapour. Let us assume that all of the three planets Venus, Earth and Mars started their lives as bodies without an atmosphere, at 0.72, 1.00 and 1.52 AU, respectively (AU = astronomical unit, the normalized distance of a planet from the Sun; Earth is, by definition, 1.00 AU from the Sun). Let us further assume that, at this stage, the geological cycle was organized in such a way that volcanoes emitted large quantities of water vapour and carbon dioxide. Now, water vapour, like carbon dioxide and methane, is a strong greenhouse gas, absorbing radiation in the infrared waveband. If there is enough water vapour and carbon dioxide to create a greenhouse effect, the planet's surface temperature will increase. It is however also important to appreciate some of the peculiarities of water. The saturation water vapour pressure is the amount of water the atmosphere can hold before water starts to condense and this is a strong function of temperature. The exact relation specifying this, the Clausius–Clapeyron equation, is exponential in shape: the hotter the atmosphere, the more water the atmosphere can hold (see also Chapter 6). On Earth, for every degree of temperature increase, the atmosphere can hold 7% more water vapour. In the case of Mars, at 1.52 AU from the Sun, the starting temperature would have been low, around 220 K. Once the atmosphere contained enough water to start condensing out to ice (around 2 Pa), it would lose all the water to ice. This is shown in Figure 1.5 as the intersection between the Mars curve and the ice saturation curve. This confirms to a large extent what we currently see on Mars—a frozen desert with a thin atmosphere.

In contrast, on Earth, starting at a higher temperature of around 270 K, only 3 K below the freezing point, water could accumulate in the atmosphere at a much higher concentration before it started to condense. The slight upward trend indicates a greenhouse

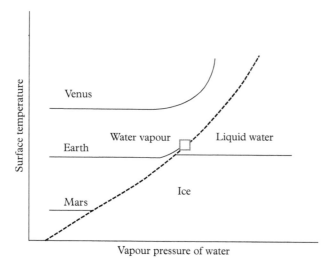

Figure 1.5 *Schematic drawing showing the runaway greenhouse effect on Venus, whereas Mars remains an ice-frozen planet, and Earth is able to hold water in its three forms. (After Kasting, 2010)*

effect caused by the water in the atmosphere. The main result is that the surface temperature increased above 273 K, allowing liquid water first to evaporate and later to condense higher in the atmosphere, where it would be colder. With virtually all water now located in the oceans, evaporation produces the counter flux, initiating a hydrological cycle on Earth. But what happened to Venus?

Venus started off warmer, at around 315 K, because of its proximity to the Sun (0.72 AU). This allowed large amounts of water vapour to accumulate in the atmosphere, but on Venus saturation was never reached. At some point, the curve goes sharply upwards and it never intersects with the saturation curve of water, as in Mars (ice) or on Earth (liquid). This is why we speak of a runaway greenhouse effect on early Venus. There are considerable simplifications in the precise application of these calculations but, by and large, they would probably have led early Venus to have a very hot and steamy atmosphere. But, without a hydrological cycle, at later stages all this water would have been lost through photodissociation (the dissolution of water into hydrogen and oxygen atoms, where the hydrogen atoms are lost to space because of their low atomic weight), leaving a hot and dry planet. On this planet, the weathering part of the long carbon cycle would have been slow, due to the absence of precipitation and because carbon dioxide emitted from volcanoes was allowed to remain in the atmosphere. There is estimated to be as much carbon dioxide in Venus's atmosphere as there is in all of Earth's carbonate rock formations.

1.6   The rise of oxygen

What happened on Earth after the temperature began to stabilize? At some point, oxygen came into the picture. The rise of atmospheric oxygen to its present-day concentration of 21% by volume serves as another good illustration of how intricately biogeochemical cycles, such as those of oxygen and carbon, interact with the geological forces and cycles of the planet. Oxygenation of the planet involves oxygenation of the main reservoirs: Earth's crust, the oceans and the atmosphere. It is commonly accepted that around 2.1 and 2.4 Gyr (1 Gyr = 10^9 years) ago, the concentration of oxygen began to rise from a minor 0.001% to its current value. This change is known as the Great Oxidation Event. However, as the range in timing already suggests, this is unlikely to have been a single big event. It is, for instance, not quite clear at precisely what time oxygen-producing bacteria entered the early planet's stage. Recent evidence suggests that this may have happened well before 3 Gyr ago. However, the current oxygenation of the deep ocean is thought not to have been fully completed until about 600 million years ago, leaving about 2 Ga to complete the oxygenation on Earth through a multitude of interactions with other biogeochemical and geological cycles.

Two chemical species play a major role in tracking the rise of oxygen on the planet: iron and sulphur. We will come back to them in Chapter 12. Their evolution is a perfect example of how geological records help us to understand Earth's biogeochemistry (Langmuir & Broecker, 2012). Today's rocks and oceans contain abundant quantities of oxidized iron and sulphur, in the forms of Fe^{3+} and S^{6+}, respectively. The oxidized form of iron, ferric iron, is non-soluble in water, whereas the reduced form, ferrous iron, is easily soluble. In contrast, sulphur's oxidized form, sulphate, is highly soluble in water, while the reduced form, sulphite, is not. This behaviour can thus be used to track Earth's oxygenation state, as the ratio of iron to sulphur in rocks indicates the level of oxygenation. An Earth with low oxygen content would appear reduced and would contain little sulphite in the oceans and large amounts of ferrous iron. Importantly, in an atmosphere containing substantial amounts of oxygen, any ferrous material exposed through weathering would become quickly oxidized and thereby lose its solubility in water. Today's oceans are thus deficient in iron, whereas current day soils, in contrast, contain more iron than older soils (>2.4 Gyr ago). This deficiency in seawater has led to a series of grand experiments to fertilize the ocean with iron to improve carbon uptake. We will come back to some of these geoengineering ideas and experiments in the final chapter (Chapter 13).

Burial of carbon can take organic carbon temporally out of the cycle, thereby shifting eqn (1.1) towards the right and so increasing the amount of free oxygen. The fraction of organic material buried, however, has remained remarkably constant over Earth's history (see Chapter 9 for more details). Nevertheless, through the use of isotopes (Chapter 3), it has been established that, around 2.0 –2.4 Gyr ago, much larger amounts of organic matter were buried. This may have led to an increased availability of oxygen. Exactly how this happened is largely unknown.

It is now assumed that the first photosynthetic bacteria appeared around 3 Ga ago. These would have produced whiffs of oxygen in an otherwise low-oxygen-content atmosphere; these are visible in the record in the early period from 2.5–3.0 Gyr ago. After this time, atmospheric levels rose, but it then took almost 2 Gyr to completely oxidize the ocean. This long time span is related to the vast size of the reservoir and the long residence times. The rates of oxygenation of Earth's ocean and atmosphere are important to understand, as they indicate the sensitivity and vulnerability to changes.

For oxygen we noted that substantial changes in these rates have occurred over Earth's history. There is an interesting corollary with climate in this story, relating to the fact that, before the oxygenation, the atmosphere probably contained large amounts of the strong greenhouse gas methane. The rise of oxygen could have oxidized this methane, resulting in a much cooler environment and starting off one of the great Snowball Earth episodes, just over 2.5 Gyr ago. However, issues about the timing and the strength of this feedback are still unresolved. Complete coverage in ice changes the reflectivity (albedo) of Earth in such a way that, how the planet escaped from this Snowball Earth episode and became warmer, is still very much an unresolved issue. It is also thought that greenhouse gases played an important role in stabilizing Earth's climate when the Sun had a much lower (70% of today's) luminosity. Without these radiative effects of greenhouse gases, Earth could have easily fallen into a much colder equilibrium (see also Chapter 10).

1.7   Non-linearity

One of the most pervasive aspects of climates and biogeochemical cycles is the non-linear behaviour of the system. Linear systems are relatively easy to understand. The reservoir analysis of mean residence time is essentially a linear problem: the flux scales linearly with the size of the reservoir (see eqn (1.4)). Furthermore, if the input changes by, say, a factor of 2, the output also changes by a factor of 2. Linear systems can be superposed: the behaviour of the systems is exactly equal to the behaviour of the sum of the parts. Non-linear systems are much harder to understand, and predicting the effects of small changes in one or two variables often requires full numerical simulation. Since Earth's biogeochemical reservoirs interact with each other, where output from one reservoir becomes input to another, each reservoir, however small, has the capacity to affect the behaviour of the whole. This crosstalk between subsystems leads to a disproportional relationship between input and output: the overall system is non-linear, and effects cannot simply be superposed. Non-linearity gives rise to unexpected events, such as abrupt transitions across thresholds; unexpected oscillations, where the frequency of the output is not equal to that of the input signal; and chaotic behaviour. Chaotic behaviour exhibits sensitivity to the initial conditions, as was established in weather forecast modelling by one of its founders, Edward Lorenz. Tiny differences in initial states can expand and even explode into large differences in later states. Importantly, the values of these variables do remain confined to fixed boundaries, while never exactly repeating the same trajectory. The Lorenz butterfly is the iconic image of this type of behaviour.

Earth system models are our current tools for understanding the non-linear interactions between climate and the biogeochemical cycles (see Chapter 3). Because of the non-linearity, they solve fundamental transport and conservation equations on a three-dimensional model grid, using large ensembles of model runs with tiny changes in initial conditions.

One final consequence of non-linearity needs to be discussed here: emergent behaviour. The climate system is not only chaotic but also complex, in the sense that myriads of interactions between its subsystems can lead to emergent behaviour that is only visible and understandable at the scale and level above the subsystems. Chaos and emergence

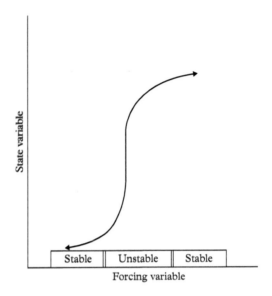

Figure 1.6 *Schematic drawing of possible stable and unstable states in a two-equilibrium system. The forcing factor (e.g. precipitation) is on the horizontal axis; the state variable (e.g. vegetation) is on the vertical axis. In this case, the stable states might be low-density desert vegetation (e.g. savanna) on the left, and a densely vegetated state (e.g. forest) on the right. In the unstable area, the system always moves into one of the two equilibrium states.*

are the two sides of the coin of complex non-linear behaviour. Figure 1.6 shows a generic example of a non-linear system. The system has two stable states, and an area in the middle where the system is unstable and moves to either one of the preferred states. Examples of this type of behaviour can be found in the vegetation of the Sahara–Sahel system, where the system moves between a desert and dense vegetation, depending on the rainfall. Another classic example is the Snowball Earth state we encountered earlier, at the beginning of Earth's oxygenation. Here, the stable states would represent an Earth covered in ice and snow, and a completely ice-free Earth, while the vertical axis would represent the energy leaving the planet. It is important to understand that, in this case, the feedback in the system works through the reflectivity dependence on temperature. Low temperatures create large reflectivities through snow and ice, while higher temperatures create low reflectivity through the absorption of solar radiation by vegetation. Another example may be Earth's oxygenation, as described earlier: either the planet is in a stable state with a considerable amount of oxygen (21%), which would be comparable to today's value, or it is in a stable state with a very low value of oxygen (0.001%). Feedbacks are thus key to the stability of the Earth system, and we will encounter many of them in the remaining chapters in this book.

   The next two chapters offer a general introduction to the context of the book, outlining the variability of climate system as a complex interplay between biogeochemistry and physics and describing some of the tools we have to understand climate: observation and models. The second part of the book consists of four chapters that introduce the climate physics necessary for understanding the interaction of climate with biogeochemistry, and climate change. These deal with radiation physics and the greenhouse effects, aerosols and the physics and dynamics of the atmosphere and the ocean. The third part of the book deals with Earth's (bio)geochemical cycles. These chapters treat the main biogeochemical cycles found on Earth: the carbon, methane, nitrogen, phosphorus, iron, sulphur and water cycles and their interactions with climate. The final two chapters bring this knowledge together to describe climate in the future and possible mitigation and adaptation actions—always with an emphasis on the interaction between biogeochemistry and climate.

2

Climate Variability, Climate Change and Earth System Sensitivity

2.1 Earth system sensitivity

This chapter is about climate, climate change and climate variability. It also deals with the sensitivity of the Earth system to changes in its forcing parameters, particularly to changes in the biogeochemical cycles. This is a story of timescales, of changing boundary or forcing conditions and of the amplification or diminishment of key processes through the operation of Earth system feedbacks.

The glossary of the American Meteorological Society defines climate as 'the slowly varying aspects of the atmosphere–hydrosphere–land surface system', it goes on to say that

> it is typically characterized in terms of suitable averages of the climate system over periods of a month or more, taking into consideration the variability in time of these averaged quantities. Climatic classifications include the spatial variation of these time-averaged variables. Beginning with the view of local climate as little more than the annual course of long-term averages of surface temperature and precipitation, the concept of climate has broadened and evolved in recent decades in response to the increased understanding of the underlying processes that determine climate and its variability. (American Meteorological Society, 2019)

Climate can also be more loosely defined as the state of the climate system, where the climate system is composed of the lithosphere, the hydrosphere, the biosphere, the atmosphere, the cryosphere and, increasingly, the anthroposphere, the human component. This wider definition also includes the dynamics and exchange of material between these spheres, through the biogeochemical cycles and their interactions—for instance, the dynamics of ice sheets. For our purposes, it is important to recognize that the state of the biogeochemical cycles is thus an integral part of the climate.

Earth is 4.54 (± 0.05) billion years old. During its lifetime, conditions both within Earth and on its surface, as well as its climate, have changed remarkably. It is important to identify the timescales and the resulting rates at which some of these changes have

Biogeochemical Cycles and Climate. Han Dolman, Oxford University Press (2019). © Han Dolman.
DOI: 10.1093/oso/9780198779308.001.0001

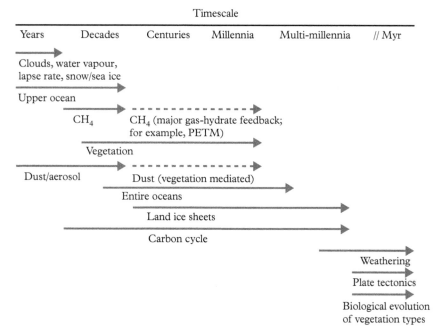

Figure 2.1 *Timescales of processes in the climate system that determine variability and feedbacks. (After Rohling et al., 2012)*

happened. Figure 2.1 shows some of the key processes determining the variability of climate (Rohling et al., 2012). Processes that are important at a million-year (10^6 year) timescale, such as tectonic processes, shown on the right-hand side of the diagram, can be considered constant at the much shorter timescales shown on the left-hand side of the diagram, as their boundary conditions do not change under shorter timescales. We have already encountered in Chapter 1 the importance of some of the slow processes, such as the geological thermostat, which involves interactions between weathering and the hydrological and geological carbon cycles. Our current climate change problem has a scale of a few centuries, essentially the time since humans started using fossil fuels. On multi-millennia to million-year timescales, the periodicity of the Pleistocene glacial–interglacial cycle is dominant.

Climate variability is defined by the World Meteorological Organization (WMO) 'as variations in the mean state and other statistics of the climate on all temporal and spatial scales, beyond individual weather events' (World Meteorological Organization, 2017). We can express the instantaneous value of a particular variable, say temperature, as the sum of a mean and a fluctuating part:

$$T = \bar{T} + T' \qquad (2.1)$$

in which $\bar{T}$ is the average value of the variable over a sufficiently long time period (sometimes taken as thirty years but, on scales of geological interest, it can be much longer), and T' is the perturbation or variation from that mean. In the case of temperature, the perturbation would be the deviation from the mean, where the mean would be derived from the temperatures over the last thirty years at a particular location. The resulting T value would be the temperature on a specific day. Any climate variable can be decomposed in this way if the time series is long enough to obtain statistically relevant values. In fact, if we take a lumped approach and consider several variables, such as temperature, humidity, pressure and precipitation, together in some way, all the values on the left-hand side of the equation would represent today's weather. The set of mean values would represent climate. Importantly, climate has always been variable and always will be variable.

Climate change, such as we experience now, or in the geological past, is a change in the mean state of the variable in eqn (2.1). This can be due to a variety of factors that we will explore in later chapters of this book. The difference between changes in the mean and those in the fluctuating part of eqn (2.1) is crucial, and finding ways to separate the two and identify forcing as opposed to internal variability is a key subject in climate science: it underlies for instance, the question of whether it is possible to attribute extreme weather events to human-made climate change.

2.2   Geological-scale variability

Let us first take a look at an example of geological-scale variability, to illustrate some key features of climate variability. In Figure 2.2, data from sediment cores taken at the ocean bottom are used to reconstruct deep-ocean temperatures (Zachos, 2008). The study used oxygen isotopes as a proxy for temperature; the technique of using these isotopes is explained in more detail in Chapter 3. For a precise explanation of the geological epochs, see http://www.stratigraphy.org/index.php/ics-chart-timescale. Figure 2.2 shows that, for the past 66 million years, the ocean temperatures have been dropping: the long-term temperature trend in the Cenozoic, comprising the periods from Palaeocene to Pliocene, is thus downward. This is a change in the mean value. However, superimposed on this trend is a large variability, as indicated by the spikes. These upward spikes are called hyper-thermals. Examples are the early Eocene Climatic Optimum (52 Myr ago) and the Palaeocene–Eocene Thermal Maximum (56 Myr ago). The latter transition is thought to have caused a global warming of 5 °C. Understanding this rise in temperature and its interaction with the suggested large increase in carbon dioxide is a very active research area, since this may provide insight into how the Earth system deals with large injections of carbon dioxide into the atmosphere and may shed light on Earth's current situation. A recent estimate puts the emission of carbon dioxide during this period at about 10% of our current emissions from fossil fuel burning (~ 10 Pg C yr^{-1}) but over a much longer time period of 4000 years (Zeebe et al., 2016). We will further explore the Palaeocene–Eocene Thermal Maximum in Chapter 9.

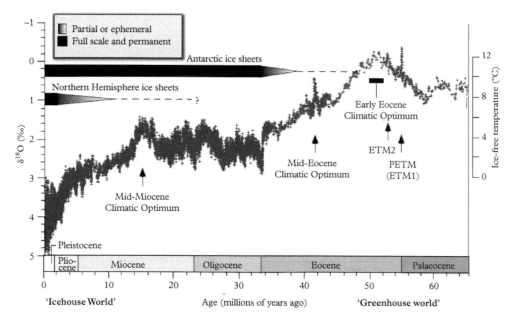

Figure 2.2 $\delta^{18}O$ *variability in ocean sediments over geological time. The* $\delta^{18}O$ *variability is a proxy for temperature shown on the right axis; ETM1, Eocene Thermal Maximum 1; ETM2, Eocene Thermal Maximum 2; PETM, Palaeocene–Eocene Thermal Maximum. (After Zachos et al., 2008)*

Most of the Cenozoic temperature variability appears related to carbon dioxide, coming from tectonic or volcanic origins. For instance, the rise around 60–50 Myr ago appears to be caused by the subduction of the carbonate crust from the Indian subcontinent, as this led to volcanic emissions that raised the carbon dioxide levels above 1000 ppm. In later stages, the concentrations of carbon dioxide declined, as the Indian and Atlantic oceans became sinks for carbon dioxide, and tectonic activity declined. Figure 2.2 also reveals the large variability in temperature during the Pliocene and Pleistocene, the period when Earth experienced glacials and interglacials. Overall, one can argue that the temperature trend suggests that Earth was slowly moving from a tropical greenhouse into a refrigerator world. The occurrence of ice sheets also appears to indicate this. These examples show that determining the average in eqn (2.1) clearly depends on the chosen timescale of interest (see Figure 2.1).

2.3 Glacial–interglacial variability

A palaeoclimatological proxy is a measurable quantity, such as tree rings, fossil pollen, ocean sediments, coral, the composition of trapped atmosphere in ice cores, and historical data that allows reconstruction of climate variability. Figure 2.2 shows that the $\delta^{18}O$ variability in ocean sediments was large in the Pliocene. However, the interpretation of the signal as a proxy for temperature is complicated by the existence of large ice sheets that also affect the $\delta^{18}O$ variability (see Chapter 8). We can also use other techniques (see Chapter 3) to investigate the variability of temperature during these cycles in more detail. In this case, we have excellent high time-resolution data from Antarctic ice cores that take us back some 800,000 years (Augustin et al., 2004; Petit et al., 1999). These ice cores captured the atmosphere when the snow compacted and turned to ice. Laboratory analysis of the air trapped in these cores provides primary data of past atmospheric composition (see Chapter 3). Figure 2.3 shows the temperature anomaly as obtained from deuterium analysis (δD depends on the ratio of ^2H to ^1H) of the ice in the Antarctic Dome C ice core, as well as combined carbon dioxide concentration data from two other cores in Antarctica (Taylor Dome and Vostok), both of which were produced by the European Project for Ice Coring in Antarctica (e.g. Galbraith & Eggleston, 2017).

This beautiful picture illustrates some key aspects of the climate–carbon dioxide interaction over the last million years or so. In contrast to the variability of the Palaeocene–Eocene Thermal Maximum seen in Figure 2.2, the variability in temperature during glacials and interglacials (often expressed as marine isotope stages) is driven by changes in the so-called Milankovitch cycles. These cycles of how Earth orbits the Sun, change the amount of incoming radiation received at Earth's surface, particularly in the Northern Hemisphere (see Chapter 6, Figure 6.8). However, as we can see from the co-variation of the carbon dioxide and temperature record, there is more to this story. In cold periods, when the ice extended well into the continents, the carbon dioxide concentration never fell below 180 ppm and, during the warmer interglacials, it never rose above 280 ppm (In 2016 the global average atmospheric carbon dioxide concentration reached 400 ppm!).

What is important is that the main timescale of this variation during the last 0.9 Myr is about 100 kyr; in the Pliocene, the timescale is estimated to be 40 kyr. How Earth transited from this 40 kyr world into the 100 kyr world is still not well established.

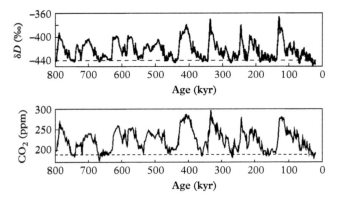

Figure 2.3 *(Top) Deuterium analysis from Antarctic ice core data as a proxy for temperature. (Bottom) Carbon dioxide variability from Antarctic ice cores. (After Galbraith & Eggleston, 2017)*

What can also be seen from the steep increase into warmer periods (high δD) is that the ice sheets break down faster than they build up. This is the familiar story of non-linearity in Earth system processes. Overall, the current theory suggests that an interglacial period will end when the summer insolation to the Northern Hemisphere reaches a minimum level. The ice caps will continue to grow, leading to overall lower atmospheric temperatures that will reduce the ocean's capability to exhume carbon dioxide, particularly in the southern oceans (see Chapter 9 for more details). Then, the overall atmospheric carbon dioxide levels will start to fall. Lower carbon dioxide will provide a further boost to low temperatures, as a positive feedback furthering the growth of the ice sheets. In time, a threshold stability of the ice sheets may be crossed that leads to disintegration of the ice sheets. A new maximum in insolation and the resulting increased carbon dioxide released from the ocean because of higher temperatures then pushes the Earth system back into an interglacial. This simplified picture is fraught with complications and details that are the subject of ongoing research but, for our discussion here, it is enough to remember that the climate state favoured now is a move towards cooler conditions (see Figure 2.2) and that the last 0.9 Myr have known a 100 kyr cycle alternating between more ice and more ice-free conditions. This fluctuation is in line with corresponding variations in the global carbon dioxide levels. How exactly this carbon dioxide level, and the level of methane, affect the temperature variability or, in contrast, how the carbon dioxide and methane concentrations vary in response to temperature are puzzles that remain to be solved and to which we will return at various stages in this book.

2.4   Centennial-scale variability

Let us now make another step forward in time and examine the variability in surface temperature over the last 100–300 years, as illustrated by Figure 2.4. This figure shows global averaged surface temperature anomalies as obtained from the instrumental record (see Chapter 3). The anomalies are calculated using a version of eqn (2.1), with the mean determined over 1951–80. So, unlike before, when we had to use proxies such as oxygen isotopes to determine the temperature, we now use real thermometers. The graph shows anomalies rather than absolute temperatures, and with good reason: absolute temperatures vary significantly from place to place and are hard to determine precisely at the global level. Anomalies, though, vary considerably less and therefore show a more realistic picture when one is interested primarily in the change over time. The data shown here are from the NASA Goddard Institute for Space Studies time series, but the four other existing global data sets all show similar patterns, with relatively minor differences. At the interannual timescale, year-by-year globally averaged temperature can vary by as much as 0.2–0.3 °C. Earth has warmed by about 1 °C over the last 150 years. This is largely due to the rise in atmospheric carbon dioxide as a result of burning fossil fuels and deforestation. This rise is shown in Figure 2.5, with the data up until 1958 stemming from ice cores and glacier firn air, and the data after 1958 being the famous series of carbon dioxide data, from in situ observations of air at the Mauna Loa Observatory on Hawaii (see Chapter 3).

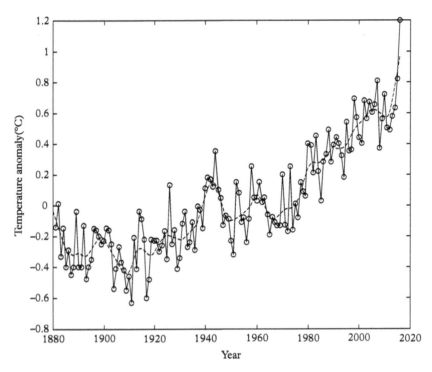

Figure 2.4 *The increase in global mean annual surface air temperature since 1880; the dashed line is based on a simple five-year moving average. (Based on http://data.giss.nasa.gov/gistemp/graphs/, accessed 31 January 2018)*

There is a difference in the structure of the variability between the temperature and carbon dioxide curves that is not just due to the way they are plotted here. The carbon dioxide curve shows a relatively smooth, almost perfect exponential increase, while the temperature curve shows plateaus and spikes going below and above the average. The carbon dioxide curve is driven mostly by exponentially increasing fossil fuel emissions, of which about 45% remain in the atmosphere, with the remainder being taken up by land and ocean (see Chapter 9). A small plateau is visible around 1940–45 as a result of reduced economic activity during the Second World War. The temperature curve, in contrast, is more variable and less homogeneous, while also showing an upward trend in the data.

The variability in the temperature data is composed of two kinds: external variability and internal variability. It is important to distinguish between these two types. Internal variability is defined as the variability that arises from the complexity of the climate system. The year-to-year variability of temperature and precipitation over a relatively small time frame is usually internal variability. Variability in ocean temperatures, and variability caused by the El Niño–La Niña cycle, belong to this type. The overall rise in temperature, as forced through the increase in carbon dioxide concentration, is external. Distinguishing between internal variability and external variability is at the heart of climate science.

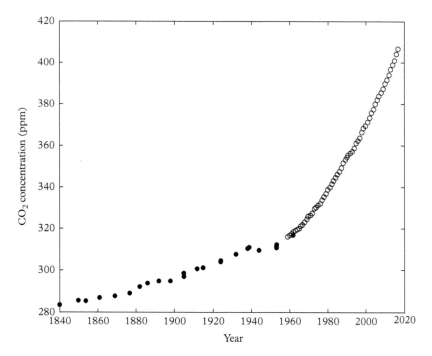

Figure 2.5 *The increase in carbon dioxide from ice core data (filled circles) and, from 1958 onwards, from observations at Mona Loa Observatory (open circles). Mona Loa data from Dr. Pieter Tans, NOAA/ESRL (www.esrl.noaa.gov/gmd/ccgg/trends/) and Dr. Ralph Keeling, Scripps Institution of Oceanography (scrippsco2.ucsd.edu/). Data accessed at 31-1-2018 at ftp://aftp.cmdl.noaa.gov/products/ trends/co2/co2_annmean_mlo.txt. Data from Law Dome DE08 from Etheridge et al., 1998)*

What causes the variability we see in Figure 2.4? Candidates for this type of forcing have been identified as greenhouse gas emissions and the associated changes in radiative forcing (see Chapter 4), solar variability, volcanic explosions causing widespread cooling through aerosols (see Chapter 5), and two large-scale ocean oscillations (see Chapter 7): the El Niño–Southern Oscillation and the Atlantic Meridional Oscillation. In particular, the ocean–land interactions operate at scales of about ten to twenty years and may have a significant influence on the trends at those timescales. However, with multiple regression techniques, it is possible to unravel the relative importance of these factors. These factors are able to explain up to 90% of the temperature variability of the last hundred years or so, suggesting strongly that the dominant cause of the trend is the rise in greenhouse gases.

2.5   Earth system variability

It is appropriate to formalize the above discussion, so that we can better understand what is happening in these cases where Earth's climate is changing. We do this by introducing the concept of climate sensitivity, S:

$$S = \frac{\Delta T}{\Delta R} \qquad (2.2)$$

Here, the climate sensitivity, following a recent suggestion by a group of palaeoclimatologists, is described as the change in temperature, ΔT, that results from a change in the radiative forcing, ΔR (Rohling et al., 2012). This definition is somewhat different from that previously used by the IPCC when climate sensitivity was defined as the change in temperature following a doubling of carbon dioxide. Equilibrium sensitivity is the change in mean global temperature after the climate system has reached a new equilibrium in response to a change in forcing, such as a doubling of carbon dioxide. Transient climate sensitivity is the sensitivity of the global mean temperature to a transient change in forcing, for instance a 1% increase in carbon dioxide per year. The global mean temperature used then is an average over twenty years since doubling of the carbon dioxide concentration is achieved.

It is important to fully understand climate sensitivity and have a definition which is not dependent on a particular period or model; otherwise, the values cannot be compared. As we have seen from the above analysis of geological-scale climate variability, the timescale of S depends on the timescale of the forcing we are interested in. This can be changes in plate tectonics, changes in weathering cycles, changes in ocean circulation, changes in Earth's albedo or changes that have occurred in more recent times, such as the changing patterns of burning fossil fuel since the Industrial Revolution (see Figure 2.1).

Figure 2.6 shows the main feedback and control loops involved in the interactions of biogeochemical cycles with climate (e.g. Goosse, 2015). In this picture, the role of radiative forcing is fundamental. Two feedbacks can be identified: (i) an internal climate system feedback loop, where we need to consider processes such as ice–albedo and cloud feedbacks, and (ii) a biogeochemical loop where increases in temperature affect key processes in biogeochemical cycling. Here, we can think of weathering cycles, but also of changes in carbon dioxide uptake and release by sensitivity of terrestrial and oceanic processes to temperature. The climate sensitivity defined in eqn (2.2) lumps

Biogeochemical cycle climate interactions

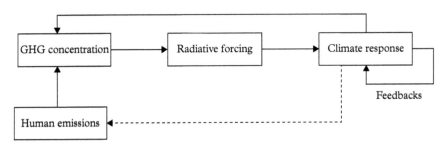

Figure 2.6 *Feedback relations causing climate variability, showing the relation between climate variability, biogeochemistry and human emissions of fossil fuel; GHG, greenhouse gas. The broken line suggests potential for mitigation and adaptation by humans to a changing climate (see Chapter 13).*

these together into a single response, focussing on how a change in radiative forcing causes a change in temperature.

Climate sensitivity thus depends on the timescale chosen. As processes that influence climate operate on many different timescales, from seconds to millions of years, it becomes important to distinguish between the timescale of the feedbacks involved. For practical purposes, a division into slow and fast feedbacks is a good choice. The definition of a slow versus a fast feedback is dependent on the time it takes for the system to reach a steady state (see Chapter 1). If a particular process results in a temperature change that reaches a steady state slower than the timescale of the corresponding radiative forcing, then we consider it a slow feedback, and the reverse for a fast feedback. An example of a fast feedback is the radiative forcing exerted by clouds, water vapour and snow cover. These processes are generally taken into account in Earth system models, albeit to various degrees of realism. If we consider the growth of ice sheets discussed above, we can call that a slow feedback. This definition inevitably contains some subjectivity, which not only relates to the processes involved but also to the resolution of, in particular, the palaeorecords. If the resolution is of a timescale larger than that of the variability, we cannot detect this variability at all. Around a hundred years is an appropriate timescale to distinguish between slow and fast processes. Taking the approach of eqn (2.2) a step further, we can define the actual (present-day) equilibrium climate sensitivity, S^a, as

$$S^a = \frac{\Delta T}{\Delta R} = \frac{-1}{\lambda_p + \sum_{i=1}^{n} \lambda_i^f} \tag{2.3}$$

where λ_p and λ_i^f are feedback parameters: λ_p is the so-called Planck feedback parameter, equal to $0.3\ \mathrm{W\ m^{-2}\ K^{-1}}$ for our present climate (we will explore the background of this value in Chapter 4; for now, it is sufficient to know that it is the longwave radiation feedback from a pure blackbody), and $\sum \lambda_i^f$ is the sum of n fast-feedback parameters. In reality, this is not the equilibrium situation the Earth system strives for; rather it strives for an equilibrium that also depends on the slow feedbacks. Incorporating now these m slow feedbacks λ_i^s as well as the fast ones yields a new equation:

$$S^a = S^p \left(1 + \frac{\sum_{i=1}^{m} \lambda_i^s}{\lambda_p + \sum_{i=1}^{n} \lambda_i^f} \right) \tag{2.4}$$

where S^p is palaeoclimate sensitivity (Rohling et al., 2012).

One may wonder why this extended treatment is necessary. The reason is that trying to calculate the response of the Earth system to human perturbation such as excessive carbon dioxide loading of the atmosphere at timescales of less than a hundred years requires insight into which factors can be considered constant and which ones cannot. Figure 2.7 shows the response of temperature to carbon dioxide forcing as obtained from a large compilation of both palaeoclimate data and model data. The slow and fast

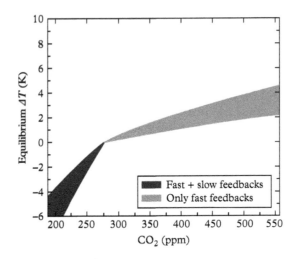

Figure 2.7 *Equilibrium response of the global temperature as a function of carbon dioxide concentration; data from the late Pleistocene of the past 800 kyr; p.p.m.v., parts per million volume. (After Rohling et al., 2012)*

feedbacks operate together in the palaeoclimate data, which are based primarily on Pleistocene records, and this is visible up to the 280 ppm maximum we encountered earlier in this epoch. On this basis, it is possible to calculate an equilibrium climate sensitivity for the doubling of carbon dioxide of 3.3–3.7 W m^{-2} K^{-1}, with a fast-feedback climate sensitivity of 0.8–1.0 W m^{-2} K^{-1}.

Over the past 65 million years, the approach yields a climate sensitivity of 0.6–1.3 W m^{-2} K^{-1} (or a slightly wider range, depending on the uncertainty). Note that the black, early part of this graph is based on observations of the variability in carbon dioxide and temperature in the Pleistocene, while the fast feedbacks, which are of major concern in adapting to current climate change, are extrapolated using eqns (2.2) and (2.3). This shows not only that the current rates of change of climate and carbon dioxide concentration have no past analogue in the recent geological past but also that estimates of future change become more uncertain at higher carbon dioxide concentrations.

For modern climate change, it becomes important to distinguish the internal variability of the climate system from its external forcing. The technique most often used for this assumes that the final forcing can be expressed as the linear sum of the various forcings. Three of these forcings are shown in Figure 2.8: the forcing from carbon dioxide and other greenhouse gases (primarily methane and nitrous oxide) following a linear increase, an IPCC scenario and the observed values (Figure 2.8(a)) and solar forcing and the forcing from volcanic aerosols (Figure 2.8(b)).

Using multiple linear regression techniques, these forcings are combined with those of El Niño (Figure 2.8(c)) and used to explain the variability observed in Figure 2.4 (van der Werf & Dolman, 2014). The residual of this model with the temperature data

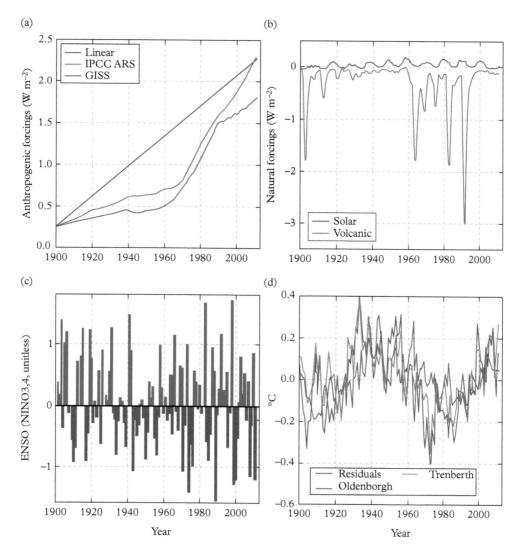

Figure 2.8 *(a) Anthropogenic greenhouse gas forcing from a linear scenario (Linear), from the IPCC Fifth Assessment Report (IPCC AR5) and from observations from the NASA Goddard Institute for Space Studies (GISS). (b) Solar and volcanic aerosol forcing (from IPCC AR5). (c) The El Niño–Southern Oscillation (ENSO) variation, as determined via the NINO3.4 index. (d) The residuals of the multiple regression plotted with two other estimates of the Atlantic Meridional Oscillation. (After van der Werf & Dolman, 2014)*

is the unexplained variability, which, in this case, is largely attributed to the Atlantic Meridional Oscillation (see Chapter 7). Figure 2.8(d) shows the comparison of the residuals with two other estimates of the Atlantic Meridional Oscillation, the agreement suggesting indeed that a large part of the variability of the residual agrees with the Atlantic Meridional Oscillation estimates. The authors of this paper showed a transient climate response over the last thirty years of 1.6 °C, with considerable uncertainty (1.0–3.3 °C). This is within the fast response range of Figure 2.7.

We have seen that climate (temperature) exhibits great variability over a large range of timescales. It is important to distinguish between the fast and slow responses of the climate system to changes in forcing and internal variability originating from the non-linearity and chaotic nature of system. It is possible using experimental—both palaeo data and modern—measurements to determine the sensitivity or climate response.

3

Biogeochemistry and Climate

The Tools

A variety of methods and tools are available to study the interactions of climate and the biogeochemical cycles. While it is not possible to give a complete overview of all available methods, this chapter mentions some of the most important ones. Because models have become increasingly powerful tools for studying both the past and the future, we also need to be aware of how they developed and what they can and cannot do.

3.1 Climate and biogeochemical observations and proxies

The instrument-based record of climate observations goes back only a few hundred years. Our first accurate thermometer dates from 1654. The first-known sea level observations were made around 1700, in Amsterdam. In 1643 Evangelista Torricelli made the first measurements of atmospheric pressure with a mercury barometer. What characterized all these different types of observations is that they provided a direct measurement of a particular climate variable, mostly at a local scale. The first rainfall measurements were taken much earlier, in 500 BC in Greece, and in 400 BC in India. The first global map of temperature is attributed to the great German explorer and naturalist Alexander von Humboldt and dates from 1817.

Networks of observations as we currently have for weather observations are something far more recent. These took off when national meteorological agencies and the WMO (which, prior to 1950, was called the International Meteorological Organization) were formed towards the end of the nineteenth century. Since then, they have played a pivotal role in improving weather forecasts and our understanding of climate variability at decadal and interannual timescales. Importantly, because these networks exist, it is now possible, albeit not without significant controversy, to derive global estimates of variability (anomalies) in climate (e.g. see Chapter 2, Figure 2.4).

In 1958 the United States launched the world's first environmental satellite, Explorer I. The first weather satellite followed two years later. Since these early days, a large number of satellites have been launched that observe climate and biogeochemical variables. In principle, satellites can measure solar energy at the top of the atmosphere to determine

Biogeochemical Cycles and Climate. Han Dolman, Oxford University Press (2019). © Han Dolman.
DOI: 10.1093/oso/9780198779308.001.0001

how much is reflected and radiated back to space by clouds; the surface temperatures and salinity of the oceans; the heights of the oceans and continents and ice sheets; and changes in gravity that can be related to changes in water and ice storage. They can also look down at Earth to expand surveillance of weather systems. Increasingly, satellites are also used to determine the chemical composition of the atmosphere. Earth observation has thus become a key tool in assessing climate change and monitoring its impacts; it provides a larger, often global, coverage than the conventional surface-based network, but often at the cost of measurement accuracy.

Charles David Keeling made the first reliable routine atmospheric carbon dioxide observations in 1956 in Pasadena, California. He subsequently transferred his system to Mauna Loa in Hawaii to start the measurements (Keeling, 1960) for what has become known as the Keeling curve (Figure 2.5). A diverse data set of snapshot measurements already existed at the start of the twentieth century, including some measurements of the 'background' concentration made from balloons and at mountain observatories. However, the importance of continuous measurements was not fully realized by funding agencies until climate change became an issue in the late 1980s. Since then, the network of atmospheric observations has steadily increased and, aided by improvements in measurement technology, now measures many more chemical species that impact climate, including reactive species, such as ozone, and aerosols.

In situ observations of the ocean are inherently more difficult, as observations of ocean variables such as temperature and salinity need to be taken from buoys or ships. For hundreds of years, sailors of both navy ships and merchant ships recorded the weather in their logbooks. In 1853 a common system of describing and archiving these observations was introduced, so that they could be used for the benefit of all shipping. Since then, of course, ocean observations have greatly expanded and probes have been developed that can be dropped off research and merchant vessels. However, considering the size of the ocean, which covers 70% of the globe's surface, the sampling density of these measurements was always very low. This changed in 1999, when ocean scientists developed a revolutionary plan, ARGO (e.g. Riser et al., 2016), to use robots to observe the pressure, salinity and temperature of the ocean. These probes dive up to 2000 m deep, resurfacing every ten days to send their data to a satellite. This observing system has been so successful that scientists now have an order of magnitude better understanding of the role of the ocean in Earth's heat balance, and of the ocean's currents and flow patterns.

Most of these methods determine the abundance of a particular atmospheric or oceanic chemical variable, or a measurement of the state of the atmosphere or surface ocean, such as temperature. Exchange of material, momentum and energy is, however, something different, and often in climate-biogeochemistry studies we are interested in the flux of material, heat or energy, rather than its concentration. In the last few decades, it has become possible to measure some of these fluxes directly with the so-called eddy-covariance method. Using this micrometeorological technique, it is possible to routinely monitor the fluxes of carbon dioxide, some other trace gases, heat, momentum and water between the land surface and the atmosphere. A global network of these sites, called FLUXNET, has been created (Baldocchi, 2014). This network is capable of monitoring the breathing of the biosphere in almost real time.

Improvements in analytical chemistry have made it possible to detect geochemical variations in Earth's geological patterns. Methods to determine chemical species at extremely small concentrations have dramatically improved, so that now a whole variety of geochemical analysis techniques are available for our purpose. The application of isotope chemistry has also led to a much greater understanding of the processes contributing to climate variability. For our purpose, accurate determination of this variability at high time resolution allows us to take a detailed look into the time history of Earth's biogeochemical evolution. This is important if we want to look beyond the current climate and investigate our geological past.

The absence of any systematic climate observations, let alone fluxes, before 1880 necessitates the use of proxies if we want to quantify and understand past climates. As mentioned before, proxy climate data are best defined as any type of preserved physical or chemical quantity or entity that can be related to a particular climate variable in the past. Proxies allow us to reconstruct climate in the same way an archaeologist uses preserved fragments to reconstruct living conditions of the past. Examples of proxies that we will discuss are ice cores, tree rings, pollen, speleothems and ocean and lake sediments. The so-called palaeothermometers are a special type of proxy, of which TEX_{86} and $U^{K'}_{37}$ are the most well known for sea-surface temperature. They are based on organic components—lipids or alkenones—in bacterial or algae membranes for which a relationship exists between the growth of the bacteria and temperature. Because these compounds are resistant to decomposition, they can even be found in ocean sediments several tens of millions of years old.

Observations of climate and biogeochemical cycles over the different domains of the Earth system are derived from a wide range of techniques. To describe all of these would require a series of books. This chapter, therefore, describes only some of the key techniques that are used in further chapters.

3.2 Physical climate observations

Let us start by introducing in situ climate observations. They form the backbone of our understanding of present climate and climate variability over the instrumental period (usually 1850 to the present). These climate records or time series are generally based on long-term meteorological observations. They include air temperature, precipitation, radiation, humidity, surface pressure, cloud cover and a set of more hydrological variables, such as soil moisture and evaporation. In the atmosphere, these measurements are often required and taken at different altitudes, using the upper-air radiosonde network.

Critical issues when using meteorological equipment are exposure, accuracy and the precision of the measurements. These issues are particularly important when the data are to be used, for instance, for detecting trends in temperature. Many of the earliest observational sites were located in rural areas that were later engulfed by urban sprawls. Because the surface properties of urban areas generally make them warmer than green rural areas, significant temperature differences can occur. How to filter out these data from the global observational data set has been a long-standing controversy.

Accuracy and precision are always problematic, and it is important to distinguish between them. Put simply, one can think of accuracy and precision in terms of a football player shooting at goal. If the player always scores a goal, even though the ball hits different areas of the net each time, he or she has a high degree of accuracy. If the player fails to score but always hits the same off-target spot, the player has a high degree of precision, but the player's team will ultimately lose through the player's lack of accuracy. Precision, or repeatability, does not always bring you to the true value you are interested in. If the player always scores, for instance, just in the higher top left corner of the net, just below the post, the player has a both high precision and accuracy. This is what we are ultimately aiming for. Good measurements are both accurate and precise. Sometimes, in atmospheric science, the terms 'bias' and 'variability' are used to describe accuracy and precision.

Uncertainties are often used in communicating results of climate change research and it is no wonder that the IPCC has devoted considerable time and effort to define these, so we follow here their analysis. Uncertainties can conveniently be classified in two types: value uncertainties and structural uncertainties. Value uncertainties arise from the incomplete determination of a particular value or result, for example, when data are inaccurate or imprecise or suffer from sampling problems (think of our station in an urban area). Structural uncertainties are altogether a completely different beast. They arise from our incomplete understanding of the processes involved, for instance when the conceptual framework or particular model used does not include all the relevant processes or relationships. By definition, they are much harder to quantify.

Value uncertainties can be estimated using statistical techniques and expressed as a probabilistic result, say, with a 90% probability level. Structural uncertainties, the 'known unknowns', can only be described subjectively, by giving the authors' single or collective judgement of the confidence in the correctness of a result. The IPCC has developed a precise series of qualifiers to achieve this. These vary from 'virtually certain' (probability >99%) to 'exceptionally unlikely' (probability <1%), with various steps in between.

Most meteorological observations are made by national meteorological services and adhere to the WMO standards. The data are collected and harvested by the WMO Information System, which is specifically designed to meet the requirements of routine data collection and automated dissemination of observed data through internet and satellite connections. This all-encompassing integrated system also forms the basis for advanced weather forecasting. Several big data centres exist where these data are available and can be found by a simple web search. Be aware of the quality of the data that you download and do not trust data sets with no description of precision and accuracy, or no published references. It is important here to note that climate data records are often based on measurements series that were not initially designed for climate purposes, the original weather data being a good example. This adds considerable uncertainty to the climate data records, as they were obtained using a system that was not originally designed for climate purposes (think, for instance, of hydrological data used primarily to assess catchment yields). Often, these data are also local observations whereas, for climate research, we often need global, large-scale data.

3.3 Climate records

Turning this diverse collection of point data sets into a global climate record is, a non-trivial exercise. Climate records are considered homogeneous if the variations in the measurements can be traced solely back to regional-scale variations of the variable of interest. Changing instrumentation, site location and the way averages are calculated all can cause inhomogeneities which need to be corrected for. Jones & Wigley (2010) mention three basic issues in the development of gridded temperature products and global and hemispheric means: (i) large-scale systematic biases (accuracy) in the data that can possibly affect the values of large areas, (ii) the lack of coverage in parts of the world, particularly before the 1950s and (iii) the homogeneity of the basic station or time series. To illustrate this, consider measurements of sea-surface temperature. These were usually made by lowering a bucket from a ship, collecting a sample and then taking the sample temperature using a thermometer. During the process of hauling in the bucket, water would be able to evaporate and thus lower the temperature. While it is possible to derive corrections, these depend on the material of which the buckets were made: wooden, canvas or, more recently, rubber. To use these data, it is necessary to determine their relative biases and to develop a history of which types of bucket were used, because different buckets were used during different times. A recent reanalysis of global temperature data did apply a new correction for these effects, with the result that previous 'observed' variability disappeared, causing considerable controversy in the field (Karl et al., 2015). Other causes of bias include exposure of the thermometer to sunshine, and the urban–rural siting mentioned earlier. The difficulty in determining all these corrections factors results in global temperature data usually being given as anomalies. Removing the biases makes these fields inherently more reliable than those of absolute values.

Because the atmosphere is well mixed, large-scale averages of temperature and pressure can be obtained using a relatively small number of sites; however, this is not the case for precipitation. For the latter, the variability is related to the scale of the rainfall producing mechanisms and system, and these can vary at the scale of tens of kilometres. For temperature and pressure, the correlation length (a measure of how the correlation between two observations declines by distance) is relatively large. For precipitation, this correlation length is small, as precipitation often falls in a very heterogeneous way.

3.4 Measuring the composition of the atmosphere

The chemical composition of the atmosphere can be measured either by collecting an air sample in a gas flask for later analysis in a laboratory or by continuous in situ monitoring. The longest record of in situ carbon dioxide concentration observations started in 1957, when Charles Keeling initiated his observations during the International Geophysical Year, 1957–8, at the South Pole and Antarctica and at the Mauna Loa Observatory, in

Hawaii (Keeling, 1960). The key discoveries from his measurements were the increasing trend, the mixing between the two hemispheres, and the seasonal cycle caused by the uptake of carbon by vegetation. This single site has now expanded to a global network of more than a hundred sites. Many of these stations now also measure methane and other greenhouse gases and related species, such as nitrous oxide and sulphur hexafluoride, as well as the oxygen-to-nitrogen ratio, and some measure the stable isotopes (see Section 3.5) of carbon dioxide and methane.

Most historical atmospheric carbon dioxide measurements have been made with so-called non-dispersive infrared gas analysers, which are based on exactly the same

Figure 3.1 *A 303 m observation tower for trace gas and aerosol measurements in central Siberia, near the town of Zottino on the Yenisei River. (Photo credit: Martin Heimann, MPI-BGC, Jena, Germany)*

principle that makes carbon dioxide a greenhouse gas: the ability of carbon dioxide to absorb infrared radiation (see Chapter 4). These instruments measure the intensity of infrared radiation passing through a 'sample' cell, relative to radiation passing through a reference cell. The differences in carbon dioxide concentrations between the sample and the reference gases allow us to determine the abundance of carbon dioxide or, more precisely, its mole fraction in dry air (in micromoles per mole, or parts per million). More recently developed instruments measure carbon dioxide with laser-based optical spectroscopic methods such as Fourier transform infrared absorption spectroscopy and high-finesse cavity absorption spectroscopy, which includes cavity ring-down spectroscopy and off-axis integrated cavity output spectroscopy, where the laser beam is put in the chamber on a slightly offset axis. These techniques are increasingly used because they require reduced calibration as a result of improved sensor stability. The target accuracy of carbon dioxide measurements is set by WMO at 0.1 ppm for the Northern Hemisphere, and 0.05 ppm for the Southern Hemisphere, while that for methane is 2 parts per billion (ppb). Reaching this accuracy requires stringent and frequent calibration against gases traceable to a common standard.

The original motivation for setting the accuracy requirements so high was to observe differences in the behaviour of continents and hemispheres. Increasingly, observations are now taken from aircraft, tall television towers or towers sometimes specifically build for atmospheric observations, as, for instance, in central Siberia and Amazonia. Figure 3.1 shows a picture of such an observation tower, which is 303 m high and located in central Siberia near the Yenisei River, a similar one exists in the Amazon forest, near Manaus.

Dealing with these different kinds of data yields problems that are similar to those described in Section 3.3 for the temperature record. Homogenization of the data, dealing with sparse networks, and biases are also important when analysing trace gas data. As part of these observations, the concentrations of stable isotopes of carbon in carbon dioxide and methane are also often measured. Isotopes, however, can also be measured not just in the atmosphere but in a whole set of materials, ranging from rocks, to shells of foraminifera (see Chapter 9). Isotopes provide key data from the past, so let us now look at isotopes.

3.5 Isotopes

Atomic nuclei of an element that contain different numbers of neutrons are called isotopes. Many elements, including carbon, oxygen and sulphur, have two or more stable, naturally occurring isotopes. Molecules containing one or more isotopes are called isotopologues. So, in the case of carbon dioxide, an isotopologue can contain isotopes of both the oxygen atom and the carbon atom. The slight difference in atomic weight caused by this gives the molecule slightly different physical properties, thus producing differences in the abundance of the particular isotope in the environment. These variations are often so small that the difference in isotopic composition is expressed in parts per thousand (abbreviated as 'per mil' or indicated by the symbol ‰) relative to a standard. In the ratio R used below, the heavier isotope is in the numerator, and the lighter is in the denominator; so, for the isotopologue ratios for the ^{13}C isotope of carbon in CO_2 and the ^{18}O isotope of oxygen in water, we obtain

$$^{13}R(CO_2) = \frac{\left[^{13}CO_2\right]}{\left[^{12}CO_2\right]} \tag{3.1a}$$

$$^{18}R(H_2O) = \frac{\left[H_2{}^{18}O\right]}{\left[H_2{}^{16}O\right]} \tag{3.1b}$$

In practice, these ratios are normalized by comparing to a standard, say, seawater in the case of oxygen isotopes in water, and the final expression is given as the symbol delta:

$$\delta\,^{18}O = \left[\frac{\left(\left[^{18}O\right]/\left[^{16}O\right]\right)_{observed}}{\left(\left[^{18}O\right]/\left[^{16}O\right]\right)_{reference}} - 1\right] \star 1000 \tag{3.2a}$$

$$\delta\,^{13}C = \left[\frac{\left(\left[^{13}C\right]/\left[^{12}C\right]\right)_{observed}}{\left(\left[^{13}C\right]/\left[^{12}C\right]\right)_{reference}} - 1\right] \star 1000 \tag{3.2b}$$

Negative delta values relate to lower abundances, and positive values to higher abundances, relative to a standard.

Most fractionation (change in a particular isotopic ratio, or delta) occurs in a mass-dependent ratio when one of the two or more isotopes of a chemical compound has a slightly modified chemical behaviour due to their mass. The classic example is carbon, where carbon dioxide with the lighter isotope ^{12}C is more effectively used in photosynthesis than carbon dioxide with the heavier ^{13}C. This is largely because the lighter isotopologue diffuses more easily through the stomata but, importantly, also because the enzyme systems involved in photosynthesis preferentially use the lighter isotopologue. Evaporation of water yields a higher abundance of the heavier ^{18}O isotope in the remaining water as the lighter isotopologue evaporates more easily. The $\delta^{18}O$ values in the calcitic shells of the unicellular protists foraminifera reflect the temperature in which they were formed, and provide, as we have seen in Chapter 2, key proxy data for temperature data in the Cenozoic.

Some elements have several isotopes; for example, sulphur has four stable isotopes. In that case, not only mass-dependent fractionation but also mass-independent fractionation, where the amount of separation does not scale in proportion to the difference in the masses of the isotopes, can occur. The latter process happens under specific photochemical conditions; one example is the mass-independent fractionation of ozone into oxygen in the stratosphere.

Radiogenic (unstable) isotopes can be used for radiometric dating—one of the most important tools in geology. Geologists use the radioactive decay of these unstable atoms in two ways: either to determine the relative amount of the decay product against the source or to use it in a similar way as described above for the stable isotopes. The first method is used to acquire precise geochronologies, while the second one determines the isotopic signatures relevant to geochemical or biogeochemical processes. The most used techniques in radiometric dating are radiocarbon ($^{14}C/^{12}C$) dating, argon–argon ($^{40}Ar/^{39}Ar$) dating (a technique that has superseded potassium–argon dating) and uranium–lead ($^{238}U/^{206}Pb$ and $^{235}U/^{207}Pb$) dating. These measurements can be made on rocks, sediments or dead or living organic material. The choice for a particular method depends fundamentally on the half-life of the radioactive isotope (the time it takes for half of the amount of radiogenic material to decay).

Traditionally, large mass spectrometers were used to measure these isotope ratios. In these instruments, the source material, after preparation, is ionized; the ion beams are then accelerated using an electric gradient and split using magnetic sources (e.g. isotope ratio mass spectrometry and thermal ionization mass spectrometry). The preparation of the material involves combustion or pyrolysing the material and purifying it by using traps, filters, catalysts and/or gas chromatography. Accelerator mass spectrometry accelerates ions to achieve further purification, thereby greatly increasing the sensitivity of isotope ratios such as $^{14}C/^{12}C$. Another technique is secondary ion mass spectrometry (abbreviated as SIMS), where the source material is bombarded with a primary ion beam to produce ions (the secondary beam) that are subsequently measured. In particular, the development of the NanoSIMS instrumentation makes it possible to do this at an extremely high spatial resolution of the proxy material. In recent years, portable high-performance laser instrumentation has made it possible to determine stable isotopic ratios in the field via cavity ring-down spectroscopy.

3.6 Ice cores

One of the game-changing observation techniques has been the study of atmospheric samples captured in ice cores (Alley, 2000). When snow falls each year, small pockets of the atmosphere become entrapped between the snowflakes. After a number of years, such snow turns to firn and later ice, with the small pockets of air increasingly immobilized until they remain trapped as small bubbles in the ice. These bubbles can be related to the atmosphere at the time of entrapment when an accurate age model of the firn and ice layer is developed that is dependent on the accumulation rate of snow. For carbon dioxide, the concentrations from ice cores from Greenland and Antarctica appear similar, indicating a very small (at least not detectable) gradient between the hemispheres in carbon dioxide before the Industrial Revolution. This is not necessarily the case for methane, where differences in interhemispheric gradients have been found and have implications for the location of the sources of methane (see Chapter 10). Ice cores have provided key data about the concentrations of carbon dioxide and methane in the Pleistocene (e.g. Figure 2.3). Isotopic data, particularly deuterium (^2H) and ^{18}O provide effective proxies for temperature. The drilling of ice cores on ice sheets such as the Greenland and Antarctic ice sheets requires many years of collaboration and effort by international groups.

3.7 Ocean sediments

In addition to classic analysis of ocean sediments for isotopes, a relatively new discipline, organic geochemistry, has provided new tools for reconstructing past temperature. The method is based on organic components—alkenones and lipids in bacterial membranes—for which a relationship exists between the growth of the bacteria and temperature. Because these compounds are resistant to decomposition, they can be found in ocean sediments several tens of millions years old. These molecules are called biomarkers. One of the best examples are special membrane lipids biosynthesized by (hyper)-thermophilic archaea: glycerol dialkyl glycerol tetraethers. This proxy has been shown to allow reconstruction of sea water temperatures over time periods up to the middle Cretaceous (Schouten et al., 2013). It is also referred to by the name of TEX_{86}, with '86' referring to the structure of the lipid molecules. Alkenones with a shorter chain length are also used, $U_{37}^{K'}$ being the best-known other example, to reconstruct past sea-surface temperature. These proxies are not without problems, one of which is caused by uncertainty about the depth and season in which the microbes were originally living. If the microbes live substantially below the surface, reconstruction of surface temperature becomes problematic; this is even more so when there are sharp gradients in the column of ocean water, such as in the tropics, which have a sharp thermocline (see Chapter 7).

We now focus our attention to another important tool, modelling. Earth system models have become increasingly advanced and are used to integrate much of what we determine experimentally. They are the key tools for understanding the past and the present and for forecasting the future evolution of climate.

3.8 Earth system modelling

Earth system models consist of a set of mathematical descriptions of interlinked processes that describe the fundamental functioning of the Earth system (Goosse, 2015). They are underpinned by the equations describing the classical laws of physics that have been known to us for a long time. These laws include the equations of motion, and the conservation of energy and matter. These equations govern exchanges between components of the system. The equations are implemented on a set of three-dimensional grids that are staggered in the atmosphere and also have several layers in the soil. Earth system modelling has become the key tool for simulating the future climate and biogeochemistry of the planet, and is used to understand transition periods in our geological past.

Earth system models owe their existence to the numerical weather forecasting models that were first developed in the 1950s. What is generally recognized as the first general circulation experiment was conducted in 1955 by Norman Philipps at the Institute for Advanced Studies in Princeton: he took on the challenge of predicting 'at least the gross features of the general circulation of the atmosphere…without having to specify the heating and cooling in great detail' (Weart, 2008). For this purpose, he split the atmosphere into two layers and a sixteen-by-seventeen grid along the circumference of a cylinder.

He showed that such a simple model could represent key aspects of the atmosphere such as a jet stream and the large circulations such as the Hadley and Walker cells (see Chapter 6). This numerical model developed over the next ten years into a full model that described essential features of the general circulation based on the fundamental equations of physics. The model included the interaction of radiation with greenhouse gases but was still extremely simple in the way it represented other features. Ten years later, the first carbon dioxide increase experiment was performed, but it would take another twenty to thirty years before these models matured sufficiently to be used in future climate prediction or the analysis of past climates. Figure 3.2 shows how, in the course of this development, more and more processes were added to the original models of atmospheric circulation. The inclusion of these processes in the models now justifies the name 'Earth system model'.

The Earth system is dynamic: it evolves over time. The typical weather patterns of the atmosphere evolve, ocean currents change and ice sheets may melt. For all these components, different timescales are relevant and need to be taken into account. Dynamical systems can be linear or non-linear. Linear systems are 'easy', in the sense that they are periodic, stable and predictable. Unfortunately, most components of the Earth system are non-linear (e.g. Chapter 1). That means that they are aperiodic, very sensitive to their initial condition and non-predictable. To make matters worse, the atmosphere is chaotic, implying that it does not always respond to forcing in an obvious way (the outcome can be different due to some small changes in the system). Forcing on the climate consist of those processes that impact climate. At a millennial timescale, this forcing may consist of changes in solar radiation (energy from the Sun), volcanic eruptions (due to the aerosols they inject into the atmosphere that can act as a cooling blanket) and the level of greenhouse gases such as carbon dioxide and methane (see Chapter 2). At longer, geological timescales, the position of the continents and ice sheets, and the change in radiation forcing through the Milankovitch cycles, become important. It is important to realize

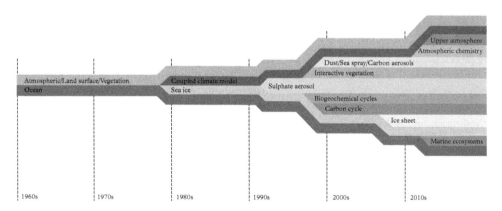

Figure 3.2 *The development of Earth system models, from early atmosphere and ocean-only models to fully complex Earth system models. (After Jakob, 2014)*

that a variable that acts as climate forcing, for instance carbon dioxide, is also an important variable in the sub-model that describes the state of a biogeochemical cycle, the carbon cycle. Forcing and model variables can thus be different or the same. To explore the behaviour of the Earth system, with its key components—atmosphere, land, cryosphere, biosphere and ocean—we require full simulation of all processes and interactions to take into account the non-linear and chaotic character.

Initial conditions are part of the general boundary conditions of an Earth system model. They describe the state of the planet at the beginning of a simulation; they should be as close as possible to the observed state. Initial conditions in chaotic systems often determine how the model simulation develops. To get insight into the climatic (long-term) properties of the model estimates, modellers use a set of model runs, each of which has slightly changed initial conditions; this then gives a set of model outcomes, which is called an ensemble. Our daily weather forecasts are also based on such ensembles, which give the forecasters a probabilistic measure of the confidence they should have in their weather predictions. Going back in time, the precise definition of initial conditions obviously becomes more and more difficult due to the lack of adequate data. Boundary conditions also vary according to the timescale of the simulation. When we run from, say, 1850 to 2050, the ocean extent, the area of ice sheets and the land-cover form are part of the boundary conditions. When we attempt a simulation of a glacial–interglacial cycle, the boundary condition of the ice sheet has to become dynamic.

Investigating the climate forcing and feedbacks that occurred during the geological past requires performing simulations of thousands to tens of thousands of years. This is difficult to achieve with state-of-the-art Earth system models, running on even the most modern supercomputers. Compromising on model resolution, complexity, number of Earth system components and the timescale of the simulation is the currently the only way to solve this. The class of model resulting from these compromises is called Earth system models of intermediate complexity (EMICs). EMICs enable us to perform the multi-millennial simulations required to simulate paleoclimate, thus helping us to understand the forcing and feedbacks that operated in the past. So-called snapshot simulations with Earth system models, where the full physics and complexity of an Earth system model is used with the boundary conditions drawn from the EMIC, are used to provide greater detail for a specific limited period.

Simplified representations of the elements of the Earth system that we have alluded to before—ocean, land, atmosphere and cryosphere—must all be included in an Earth system model. Figure 3.3 shows these elements and the key exchange processes of water, momentum, energy and chemistry. It is useful to make a distinction between dynamic processes, which describe the flow in the oceans and atmosphere, and the parameterization of physical processes such as radiation, cloud formation and land–atmosphere exchange.

The water in the ocean, and the air in the atmosphere, are fluids and their motion is described by the three-dimensional Navier–Stokes equations. These are often referred to as primitive equations. These highly non-linear equations (see Chapter 6) cannot be solved analytically; they must be solved numerically. In fact, they are listed as one of the thirteen fundamental Millennium Problems in mathematics. The equations describe the

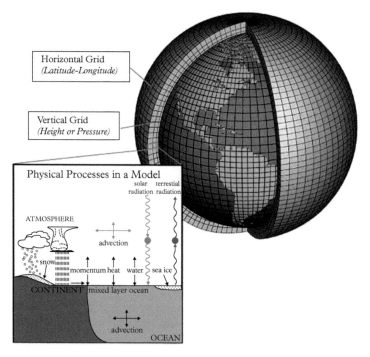

Figure 3.3 *Earth system model structure. (Reproduced with permission from http://celebrating200years.noaa.gov/breakthroughs/climate_model/modeling_schematic.html)*

relationship between the velocity, pressure, viscosity and density of a moving fluid. In a very general sense, the equations show that the acceleration of the fluid (i.e. in the ocean or in the atmosphere) is a function of gravity and pressure differences. The fact that they need to be solved on a rotating Earth adds further complexity.

The dynamical equations are solved for each point in a three-dimensional rectangular grid covering all of Earth. For each grid point, the three-dimensional wind speed, pressure, temperature and humidity, and the concentration of passive chemical species, are calculated, with the values then assumed to represent the mean conditions in the box. These variables represent the set of state variables of the model. Figure 3.4 shows how the Earth system model splits up the world into tangible (finite distance) vertical and horizontal grids. The horizontal grids are based on longitude and latitude. The vertical grids are based on pressure or height and are best thought of as one-dimensional columns composed of a series of grid boxes. While the original global circulation models, which only dealt with the atmosphere, would divide up the atmosphere into only two levels, current Earth system models typically use several tens of layers. Similarly, horizontal grid boxes have come down from being several degrees (5°–10°) long to now being 0.25° long, while some simulations can divide up the globe into 1 km grid boxes. This is important to appreciate, as the grid length determines which features a model can resolve. Typically, the scale at which a phenomenon can be resolved in a model requires the grid length to be at least half the length scale of the phenomenon. So, if we want to resolve a typical Northern Hemisphere weather system such as an anticyclone of about 200–300 km in diameter, we need grid lengths of at least 100–150 km.

Physical processes such as radiation are dealt with in more detail in Chapters 4, 5 and 6. Here, it is sufficient to say that they form an essential component of any Earth system model, because they provide the key energy input into the system. Furthermore, they describe the crucial interaction with greenhouse gases, the atmospheric part of the bio-geochemical cycles. Following Jakob's (2014) approach, building climate models involves four fundamental steps, of which expressing the fundamental laws, such as the dynamic processes, in mathematical terms and then solving them using numerical approxima-tions are the first two and refer to our physical processes. The third is an important area for research and is called parameterization. Parameterization does not aim to physically model a particular process; rather, it aims to model the effect of that process. This hap-pens for those processes that are excluded from the fundamental model equations but are nevertheless very important. It includes processes that act on scales smaller than those represented by the numerical model grid, and processes for which there are no straightforward equations or for which our understanding is incomplete, such as the majority of biological processes. Parameterization also includes the formation of clouds, and biogeochemical processes. Parameterization is a fundamental necessity in Earth sys-tem models, but also causes considerable discussion about the correctness and the resulting uncertainty. It should not come as a big surprise then that the key uncertainties of Earth system models revolve around cloud parameterizations, aerosols, precipitation and, importantly for this book, biogeochemical cycling.

Around 2000, the first Earth system model studies were published that incorporated biogeochemical processes. This was an important milestone for three reasons. First, cli-mate change alters the biogeochemical cycling of greenhouse gases such as carbon diox-ide and methane, which act as radiative forcing (see Chapter 4). Second, changes in atmospheric composition influence the biogeochemistry of radiatively active compounds. Third, climate change alters the biogeochemistry of substances that are not radiatively active in themselves but that affect the atmospheric concentration of other climatically active compounds. Most attention so far has been focussed on the first reason, because key processes that involve photosynthesis and respiration are temperature sensitive.

Earth system models suffer from our limited knowledge of the physics of some of the key processes, such as the effect of aerosols, cloud formation and precipitation, that contribute to what is known as 'structural uncertainty'. Despite tremendous increases in computing power, Earth system models still suffer from a limited ability to resolve space and time at the required detail.

3.9 Inverse modelling

The models discussed above are so-called forward running models; they predict the evolution of variables for the next time step, based on their actual state and the changes calculated from that. Another class of models often used in biogeochemical modelling

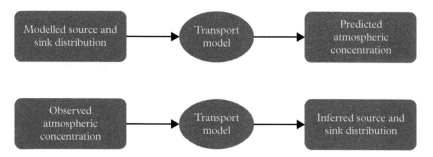

Figure 3.4 *Forward and inverse model sequences. The upper sequence is for a forward model; the lower sequence shows the inverse modelling approach, where source and sink distributions are inferred from observed concentrations.*

consists of inverse models. These models essentially run backwards in time. In biogeo-chemical modelling, they are used to infer sources and sinks of trace gases from observed concentration distributions. In Figure 3.4, the main difference between a forward mod-elling system and an inverse modelling system is shown. Note that not only source and sink distributions can be inferred; in some cases, parameter values of the models may also be estimated. The (linear) inverse problem can be written as

$$\mathbf{d} = G\mathbf{m} \tag{3.3}$$

where G is a model (in our case, usually an atmospheric transport model), d is the data vector (carbon dioxide concentrations) and m is the best estimate of sources and sinks. Inverse modelling consists of finding a set of optimal fluxes (sources and sinks) which satisfy all available pieces of information (measurements and prior estimates of the fluxes) within their respective uncertainties. In the case of so-called Bayesian inversions, a first guess ('a priori') distribution of sources and sinks is given that is then stepwise further optimized. The resulting ('a posterior') distribution with uncertainties is the answer we are looking for.

   Inverse problems are generally not easy to solve, for two reasons. First, different val-ues of the model parameters may be consistent with the data (this characteristic is also known by the term 'equifinality'). Second, discovering the values of the model parameters may require the exploration of a huge parameter space. However, when applied carefully, such models give valuable information about, for instance, the interannual variation and continental-scale distribution of sources and sinks of carbon dioxide.

4

The Physics of Radiation

4.1   Radiation first principles

The interaction of energy from the Sun with Earth and the atmosphere determines to
a large extent the climate and habitability of our planet. As shown in Chapter 1, for
water to be present in a solid, liquid and vapour form requires Earth to have a rather
narrow range of temperature variations. To understand how energy from the Sun
reaches our planet and how it is used requires us to go into some detail of the physics
of radiation and radiative transfer, because this is the main form in which energy is
received. To understand the impact which changes in the composition of the atmosphere have on our planet, we first need to understand the basic physics of radiation and
apply this knowledge to simple models of radiative transfer. The greenhouse effect and
the resulting radiative balance that is responsible for keeping Earth's temperature within
the required limits is fundamentally important for understanding our past climate and
future climate change.

Imagine a satellite in an orbit around Earth: it has radiometers on board that point
both to the Sun and to Earth. The radiometers measure not only the intensity of the
radiation but also its spectral composition. What differences are to be observed in either
quantity? Radiation consists of electromagnetic waves. These waves are characterized by
two fundamentally related properties: wavelength and energy. Figure 4.1 shows the electromagnetic spectrum and identifies the key regions of electromagnetic radiation, from
radio waves at long wavelengths (order of metres), to microwaves and infrared light, to
visible light, to the ultraviolet, X-ray (röntgen) and gamma-ray regions. The Sun emits
its energy primarily in radiation with wavelengths in the 'visible' (i.e. for humans)
domain. The visible region is generally defined as the range extending from 3.9×10^{-7} m
to 7.6×10^{-7} m. Shortwave (solar) radiation is confined to wavelengths shorter than
4×10^{-6} m and longer than 1×10^{-7} m. Longwave (infrared) radiation has wavelengths
longer than 4×10^{-6} m and up to 1×10^{-4} m. This distinction is important if we are to
appreciate what our sensors on board the satellite will observe. Shortwave radiation
dominates the radiation from the Sun, and longwave radiation dominates the outgoing
radiation from Earth. This aspect of radiation is related to the temperature of the emitting body.

Biogeochemical Cycles and Climate. Han Dolman, Oxford University Press (2019). © Han Dolman.
DOI: 10.1093/oso/9780198779308.001.0001

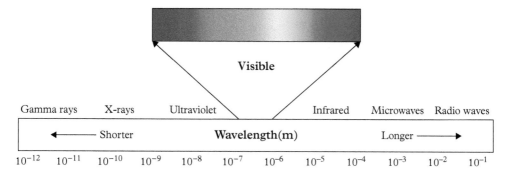

Figure 4.1 *Spectrum of electromagnetic radiation (wavelengths are in metres).*

Energy transferred by electromagnetic radiation at a particular wavelength λ per unit time, per unit area, per wavelength unit and per unit solid angle is called the monochromatic intensity, I_λ. The intensity, or radiance, I over a particular range of wavelengths is then given by the following integral:

$$I = \int_{\lambda_1}^{\lambda_2} I_\lambda \, d\lambda \tag{4.1}$$

Wavelength and frequency are, from first principles, always connected through a simple relation: $\lambda = c/v$, with c the speed of light in a vacuum, and v the frequency. So, equations like eqn (4.1) can be written in both frequency and wavelength.

An important concept in radiation physics is that of a blackbody. A blackbody is an element or surface that completely absorbs all incident radiation. An example is a piece of charcoal with a mat, dark black surface. Surfaces are blackbodies only for a particular range of wavelengths, that is, they only absorb longwave or shortwave radiation; they seldom absorb both types. Bodies in this context can be anything from solid media to gases; what is important is that they have a well-defined temperature. The radiation of a perfect blackbody is very closely approximated by a cave with a very small aperture where the small amount of incoming radiation is reflected only on the inner surface of the cave and is not returned back though the aperture.

The blackbody radiation intensity depends only on the temperature and wavelength, and it is the maximum possible intensity for any kind of body. The intensity of radiation emitted by such a blackbody (B_λ; expressed in watts per square metre per micrometre of wavelength) with absolute temperature T is given by the Planck function

$$B_\lambda(T) = \frac{2hc^2}{\lambda^5 (e^{hc/\lambda kT}) - 1} \tag{4.2}$$

where c is the speed of light in vacuum (3.00×10^8 m s^{-1}), h is Planck's constant (6.62×10^{-34} J s) and k is Boltzmann's constant (1.38×10^{-23} J K^{-1}). These are fundamental physical constants.

Figure 4.2 shows the results of calculations made using eqn (4.2) for two surfaces: hot surfaces with temperatures equivalent to the outer rim of the Sun ($T \approx 6000$ K), and surfaces more similar to that of Earth (300 K); note the different vertical scales. Our radiometers on the satellite will thus detect the high-intensity shortwave radiation from the Sun, and the low-intensity and longwave radiation from Earth. There is a factor of 10^5 difference in the intensity of the emitted radiation! Not only does the spectrum of radiation from hot bodies contain more energy, the irradiance peak also shifts towards the left (i.e. lower wavelengths) when temperature increases. The peak can be calculated by differentiating the Planck function with respect to wavelength and setting the result to zero (the equivalent of finding a maximum of a mathematical function). This yields Wien's law, the second important radiation law:

$$\lambda_m = \frac{b}{T} \tag{4.3}$$

where λ_m is the wavelength of maximum radiation emission, and b is Wien's displacement constant, which is equal to 2.897×10^{-3} m K. Using Wien's law gives us a λ_m for the Sun (remember, the outer core has a surface temperature of around 6000 K) of 5×10^{-7} m; the equivalent radiation peak from Earth (surface temperature of around 288 K) gives a peak at 1×10^{-5} m (see Figure 4.2). While the value for the Sun is in the middle of the

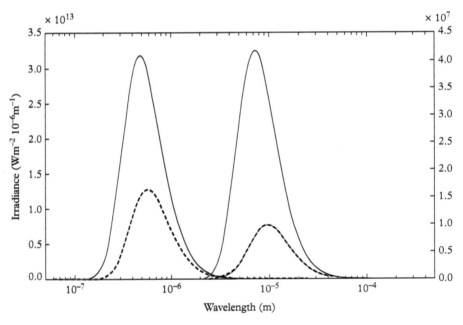

Figure 4.2 *Irradiance spectra for blackbodies calculated with the Planck function (eqn (4.2)) for hot surfaces such as the sun (6000 K, solid line; 5000 K, dashed line) on the left, with the left axis, and cooler surfaces like Earth (500 K, solid line; 300 K, dashed) on the right, with the right axis. Note the difference in scale between the left axis and the right axis!*

visible waveband, the emitted radiation from Earth is in the infrared. This change in maximum peak intensity and the associated spectrum from the Sun to Earth is called the spectral shift, and our radiometers will detect precisely this shift.

Integrating the Planck function over all wavelengths from 0 to ∞ and over all directions going out of a plane surface yields an equation for the radiation density (in watts per square metre) emitted by a blackbody provides us with us our third important radiation law:

$$F = \sigma T^4 \tag{4.4}$$

This equation is known as the Stefan–Boltzmann law. The Stefan–Boltzmann constant, σ, is 5.67×10^{-8} J m^{-2} K^{-4} s^{-1}. Equation (4.4) effectively states that doubling the temperature of a blackbody increases the rate at which it emits radiation by a factor of 16! The Stefan–Boltzmann law, together with the Planck function and Wien's law, describes the fundamental properties of blackbody radiation.

Up to now, our satellite has been observing the Sun and Earth's radiation as if there were no atmosphere. The spectral shift was based purely on the magnitude of the different surface temperatures. However, to understand the greenhouse effect, and the precise way in which the atmosphere and its greenhouse gases interact with radiation, we need to invoke our knowledge of the atmosphere.

4.2   Scattering, absorption and emission

When radiation enters a cloud-free atmosphere, it can be scattered or absorbed (and subsequently emitted). The net effect of these processes is linearly proportional to (i) the intensity of the radiation, (ii) the local concentration of gases that scatter or absorb and (iii) the effectiveness of these gases as scatterers or absorbers. There are two important types of scattering: Mie scattering, which takes place in atmospheres with relatively large particles such as aerosols and water droplets, and Rayleigh scattering, which occurs primarily with molecule-sized particles. The strength of Rayleigh scattering is proportional to $1/\lambda^4$, so the smallest wavelengths are scattered most. Thus, Rayleigh scattering gives us the characteristic blue colour of the sky as radiation with wavelengths around the blue ($\lambda \approx 0.425 \, \mu \, 10^{-6}$ m). At sunset or sunrise, the low-angled direct beam of radiation experiences a longer path through the atmosphere and therefore a larger concentration of scattering molecules. Most of the blue radiation will then be scattered out of the beam. As a result, this leaves more radiation in the reddish ($\lambda \approx 0.65 \, \mu \, 10^{-6}$ m) wavelengths. Similarly, smoke, for instance from burning, will generate large particles as aerosols and give the sky a typical reddish gloom. These aspects need not concern us now, as we are primarily interested in the radiative properties of greenhouse gases and how they contribute to long-term stable climates and their variability.

Before understanding how absorption of greenhouse gases regulates temperature in the atmosphere, we must first consider the ways in which molecules can store (and emit) energy. The first is the 'standard' energy of particles having kinetic energy through their motion in space. Molecules can also rotate and vibrate, giving them two other forms of

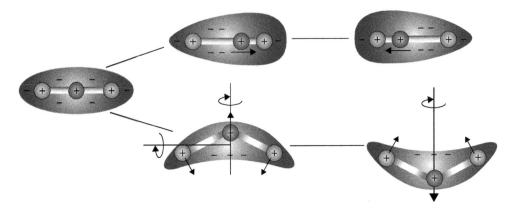

Figure 4.3 *Vibration (upper panels) and rotation modes (lower panels) of a linear symmetric triatomic molecule such as carbon dioxide, with associated charge distributions. (After Pierrehumbert, 2010)*

energy: rotational energy and vibrational energy (Figure 4.3). The second form of energy is related to changes in the energy state of electrons, which gives the energy of an emitted photon as the difference in the energy states of the electron. This form of absorption takes place mostly in the ultraviolet range (1×10^{-8} m $< \lambda < 4 \times 10^{-7}$ m) and concerns mostly molecules of oxygen, ozone and nitrogen. The absorption follows the quantum rules, in that only discrete steps between the electron energy levels are allowed. This gives rise to discrete absorption lines.

Carbon dioxide and water, however, absorb mainly through the vibrational and rotational transitions shown in Figure 4.3. This happens primarily in the infrared. This transition involves displacement of individual nuclei leading to changes in the electro-magnetic field around the molecule. The vibration creates a change in the distribution of the charge in an otherwise neutral molecule, called a dipole moment. This dipole moment plays a crucial role in the interaction with the incident radiation at the wavelength hitting the molecules. One way to think of this is that the nuclei are connected through quantum-mechanical springs, each having different spring (stretch) constants. This allows them to vibrate between the original, equilibrium state and the exited or stretched states. If they were not connected by springs but by fixed rods, they would not be able to vibrate, and the only option to absorb energy would be through rotation. Carbon dioxide is such an efficient absorber because the vibrational states and the bending states shown in Figure 4.3 fit well within the wavelengths of the infrared radiation emitted by Earth. Molecules with more than three atoms, such as methane or the chlorofluorocarbons, have more complex modes of vibration and rotation. Once the molecule, through stretching or vibration, has absorbed a photon, the additional energy becomes part of the internal energy of the molecule; this can then be released later by emission (at the same wavelength) or through collisions, thus heating the environment or the atmosphere.

Important parameters describing absorption and emission are the absorptivity a_{λ}, which is the fraction of the incident radiation that is absorbed (at wavelength λ), and the emissivity, ε_{λ}, which is emitted radiation expressed as a fraction of the blackbody radiation for the same temperature (again at wavelength λ). Both are dimensionless numbers between 0 and 1. To progress, we need to understand one more fundamental equation in radiation physics: Kirchhoff's law. Kirchhoff's law states that emissivity at a particular wavelength, ε_λ, of a body equals absorptivity at that wavelength, a_λ:

$$\varepsilon_\lambda = a_\lambda \tag{4.5}$$

To understand this deceptively simple equality, consider a body radiating inside a cave whose walls behave as a perfect blackbody (i.e. they absorb all radiation). Both the object and the walls of the cave have the same temperature: the walls and the body are in radiative equilibrium. Now, if the body absorbs a fraction a of the incident radiation (as given by the Planck function), the requirement of radiative equilibrium (same surface temperatures) implies that the body must also emit a similar fraction, ε, of that radiation. If the body were to emit more, radiative equilibrium would be broken. Note that, if the body with temperature T_{body} is placed in a different environment (with $T \neq T_{body}$), the property a_λ ($= \varepsilon_\lambda$) remains the same; in particular, the emitted radiation would be given by eqn (4.4) multiplied by the emissivity. For a blackbody, absorption equals emission at its maximum, so $a_\lambda = \varepsilon_\lambda = 1$. This also gives us a quick definition of a grey body: $a_\lambda = \varepsilon_\lambda < 1$. Kirchhoff's law also applies to gases. In our atmosphere, below altitudes of about 60 km, the conditions for Kirchhoff's law apply.

The limited set of vibrational and rotational states of greenhouse gas molecules implies that interaction of them with radiation is confined to those energy levels (at specific wavelengths) that can excite the molecule or cause it to vibrate or rotate, like a piano tuner's set of tuning forks that can only vibrate at a limited set of frequencies.

This gives rise to gases that are effective absorbers only at specific wavelengths. Absorption then occurs as spectral lines, the centre of which is located at the particular wavelength of the absorption (or emittance). Two other characteristics that are important are the width of the line and the strength of the line. The strength of the line is proportional to the probability that, during a molecule–radiation interaction, absorption takes place, while the width is partly caused by Doppler broadening—related to the phenomenon that a molecule that moves towards, or away from, a radiation source, say, the Sun, will experience a slight shift in frequency of the measured radiation—and partly caused by pressure broadening that is due to collision of molecules (and becomes important at high pressures in the atmosphere). The width of a line depends on how quickly with a characteristic time, or half-time, the absorbed radiation (at the higher energy state) can be returned as emitted radiation (inducing a lower energy state). Variations in absorption with wavelength depend strongly on the strength and spectral position of the absorption lines and how they are grouped into bands. The spectral position in relation to the Planck function for the solar spectrum is also important. If the spectral band of a particular gas that is absorbing is strong but lies in a region where the emission is low, the band is not very effective; in contrast, if a weak absorption band lies in a region where the emission is high, the effect may be large.

Carbon dioxide has a strong absorption band near 15×10^{-6} m, near the peak of the longwave radiation at atmospheric temperatures according to the Planck function. So, even

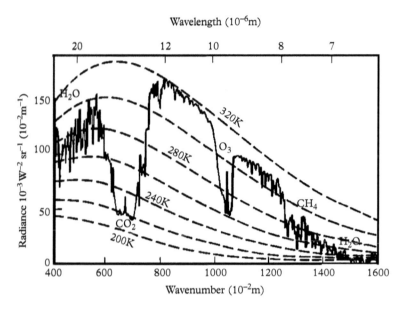

Figure 4.4 *Infrared radiation emitted from Earth's surface and atmosphere, as observed over North Africa on 5 May 1970 at 12.04 GMT from a satellite instrument orbiting above the atmosphere (solid line). Planck curves of predicted radiation from a blackbody at various temperatures are superimposed on the spectrum (dashed lines); 320 K = 46 °C; 300 K = 27 °C; 280 K = 7 °C; 260 K = –13 °C; 240 K = –33 °C; 220 K = –53 °C; 200 K = –73 °C. (After Hanel et al., 1972)*

while, as a trace gas, it constitutes at present only around 0.04% (400 ppm) of the atmosphere, it has a large effect on Earth's radiative balance. This is demonstrated in Figure 4.4, which shows atmospheric spectra from a satellite over North Africa. There is a part of the spectrum where the outgoing radiation is not absorbed: a so-called atmospheric window where the atmosphere is transparent to wavelengths from $10 \ 10^{-6}$ m to $12 \ 10^{-6}$ m. The absorption bands of carbon dioxide, ozone, methane and water are clearly visible. This graph shows how closely the absorption of carbon dioxide coincides with the peak emissions from the Planck function at the typical temperatures at which the atmosphere radiates. In short, this is what makes carbon dioxide such an important greenhouse gas on Earth. It also provides the basis for some of the greenhouse gas measurement techniques described in Chapter 3.

To make matters rather more complicated, the relationship between absorption and concentration is highly non-linear. We define absorption as 'strong' over a specific spectral interval $\Delta\lambda$ with an absorbing path length Δs if most of the energy of the incident radiation is absorbed, and as 'weak' if most of the energy is transmitted. In the weak condition, the absorption is directly proportional to the amount of absorber (or the path length Δs). However, when we deal with strong absorption, it can be shown that the absorption becomes proportional to $\sqrt{\Delta s}$ (e.g. Taylor, 2005). In this case, absorption is saturated at the centre of the band. When the concentration increases, absorption grows only at the wings of the band. Importantly, the absorption of carbon dioxide is saturated, so it takes a fourfold increase in concentration to double the absorption.

We can express the fate of a beam passing through an atmospheric layer consisting of gases as

$$\Delta I_\lambda = -I_\lambda \rho r k_\lambda \Delta s \tag{4.6}$$

where I_λ is the intensity of the beam, ρ is the density of the gas, r is the mass of the absorbing gas per unit mass of air (specific weight) and Δs is the length of the absorbing path. Here, k_λ is the absorption coefficient, in mass terms; it can be interpreted as an interception area per unit of mass of the absorbing constituent. A similar expression can also be given for scattering: in that case, the coefficient is the scattering coefficient. Thus, a beam crossing a part of the atmosphere loses intensity through scattering and absorption. In an optically thin atmosphere, the beam crosses the atmosphere and is attenuated by an amount given by the application of Beer's law. This can be obtained by integrating eqn (4.5) and, neglecting scattering for the time being, the final result becomes

$$T = e^{-\rho r k_\lambda \Delta s} \tag{4.7}$$

with T being the transmittance, the ratio of the intensity of the remaining beam to the original. An optically thin atmosphere is defined when $T \approx 1$, and an optically thick (opaque) case when $k_\lambda \Delta s \gg 1$, and $T \ll 1$.

4.3 Two-layer radiation models

Let us now apply these insights to a simplified atmosphere that consists of two layers with temperatures T_1 and T_2, respectively, and a surface with the temperature T_0 (Figure 4.5). The atmospheric layers are opaque for longwave radiation but transparent for shortwave radiation. The layers emit radiation according to the Stefan–Boltzmann law, eqn (4.4). The two atmospheric layers are in equilibrium, where the following three energy balance equations hold:

$$2\sigma T_2^4 = \sigma T_1^4 \tag{4.8a}$$

$$\sigma T_2^4 + \sigma T_0^4 = 2\sigma T_1^4 \tag{4.8b}$$

$$S + \sigma T_1^4 = \sigma T_0^4 \tag{4.8c}$$

where S is the incident radiation at the top of the atmosphere. Solving these equations gives the temperatures of the layers and the surface temperature as a function of S. Increasing the number of layers, of course, would give a more detailed approximation of the temperatures than that of this simple two-layer model. An equation for the evolution of the temperature as a function of the radiation flux divergence with height (Gaevskaya et al., 1962) is given by:

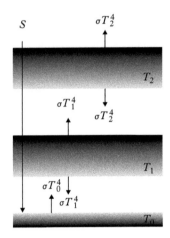

Figure 4.5 *A simple blackbody, two-layer model for calculating radiative equilibria in the atmosphere.*

$$\frac{\partial T}{\partial t} = \frac{-1}{\rho c_{\mathrm{p}}} \frac{\partial F}{\partial z} \tag{4.9}$$

where ρ is the density of the atmosphere (in kilograms per cubic metre) and c_{p} is the specific heat at constant pressure (in joules per kilogram per Kelvin). This type of equation, the continuity equation, is very common in atmospheric science and fundamental in all physics (see Chapter 6). In this derivation of radiative equilibrium, all motions of the atmosphere, such as convection, that change the temperature profile of the atmosphere are neglected. Generally, the impact of convection is to reduce the gradients in temperature, as rising air is a very effective mixer of temperature gradients (although not to zero; see Chapter 6). In the real atmosphere, the temperature structure of the troposphere, the lowest 10–15 km of the atmosphere, is influenced mainly by convection, while, in the stratosphere above, radiative impacts dominate. This contrast is directly related to the different optical densities of the troposphere and stratosphere. The stratosphere is optically thin for infrared, while the troposphere is optically thick, or opaque. The difference is important for understanding the greenhouse gas effect that is described in the next section.

4.4   The greenhouse gas effect

A simplified temperature profile of the atmosphere is given in Figure 4.6 (e.g. Taylor, 2005). It consists of a tropospheric profile in which the temperature decreases with height, and a stratospheric profile with no gradient, that is, where temperature is uniform with height. While the real temperature profile is more complicated, this simplification is adequate for our purposes here. (In Chapter 6, more details are presented on the real structure of the atmosphere.) If the optical density becomes large, the atmosphere is

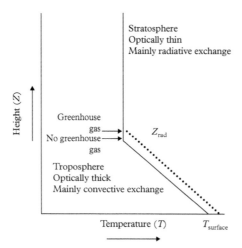

Figure 4.6 *Simplified representation of the atmosphere showing the different tropospheric and stratospheric temperature gradients.*

called 'thick', or opaque for a particular wavelength. If it is close to zero, all radiation of that wavelength is transmitted without being absorbed. Now let us assume that we add a finite amount of a greenhouse gas, such as carbon dioxide, which absorbs in the infrared. If we further assume that the atmosphere consists of a number of slabs, each of which acts like a blackbody, as in Figure 4.5—that is, that the emitted radiation follows the Planck function at a particular temperature—it follows that they are also perfect absorbers. Infrared radiation can then only be emitted from the uppermost slab, near the tropopause (Figure 4.6), which is at the temperature defined as T_{eff}. The height at which this occurs is generally thought of as the effective radiative level of the atmosphere, z_{eff}. The lower levels at temperature T_z radiate at σT_z^4, while the effective radiation level radiates longwave radiation into space at σT_{eff}^4. This is the radiation our satellite radiometer observes when looking towards Earth. The effect of the carbon dioxide absorption band is evident also in Figure 4.4, which shows that net radiation in space in that band only occurs at heights where the temperature is below −220 K.

If the greenhouse gas concentrations grow, this causes more absorption of radiation in the upper layers of the longwave radiation coming from lower layers and, at the same time, more emission from these layers to space. If the temperature change is neglected for the moment, the additional absorbed radiation in the upper layers, denoted by subscript 'up', is $\Delta\varepsilon_{\text{up}}\sigma T_0$, with T_0 being the temperature at the surface. The additional emission to space is then given by $\Delta\varepsilon_{\text{up}}\sigma T_{\text{up}}^4$. Consequently, the total terrestrial emission to space is changed by $\Delta\varepsilon_{\text{up}}\sigma(T_{\text{up}}^4 - T_0^4)$, which is negative because of the permanent temperature gradient. This causes the total outgoing radiation to drop below its equilibrium with the absorbed solar radiation at the surface. This leads, in turn, to the troposphere warming up until the balance is restored. So, the lower atmosphere ends up with permanently increased temperature: this is the greenhouse effect. As seen, the vertical gradient of T

when Earth's surface is warmer than the overlying atmosphere, is an essential part of the origin of the greenhouse effect.

What would be the temperature at which this effective outgoing radiation occurs? To answer this question, we develop a very simple model of the radiation balance of Earth, or of any planet receiving radiation from a nearby sun (Figure 4.7). The Stefan–Boltzmann law (eqn (4.4)) can then be applied to the Sun with an outer rim temperature, T_{sun}, of 5800 K and a radius, R_{sun}, of 7×10^8 m:

$$P_s = 4\pi R_{sun}^2 \sigma T_{sun}^4 \tag{4.10}$$

where P_s for power (in watts) is the total emitted radiation of the Sun. The radiation crossing a sphere with radius equal to the distance $d = 1.5 \times 10^{11}$ m from the Sun to Earth's centre is given by the middle term of eqn (4.11). Substitution of eqn (4.10) into the middle term of eqn (4.11) then yields the radiation at the top of Earth's atmosphere, F_{TOA}, also called the solar constant, which is 1392 W m^{-2}:

$$F_{TOA} = \frac{P_s}{4\pi d^2} = \frac{\sigma T_{sun}^4 R_{sun}^2}{d^2} \tag{4.11}$$

From observations, this value is closer to 1370 W m^{-2} but this is close enough for our purposes. The total energy radiating from Earth, P_{Earth}, is given also by the Stefan–Boltzmann law, but now evaluated with R_{Earth}, the radius of Earth, and T_{Earth}, its radiative temperature:

$$P_{Earth} = 4\pi R_{Earth}^2 \sigma T_{Earth}^4 \tag{4.12}$$

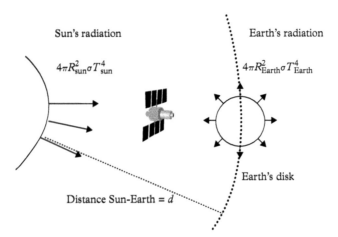

Figure 4.7 *Schematic drawing of the sun emitting radiation that is received by Earth. Note that Earth's disc has an intercepting area of πR_{Earth}^2. Also shown is our satellite pointing towards the Earth.*

This amount of energy must be balanced by the part of the incoming energy that is intercepted by Earth, $\pi R^2_{\text{Earth}} F_{\text{TOA}}$, minus the fraction α of the solar radiation being reflected (the planetary albedo):

$$P_{\text{Earth}} = (1-\alpha)F_{\text{TOA}}\pi R^2_{\text{Earth}} = 4\pi R^2_{\text{Earth}}\sigma T^4_{\text{Earth}} \qquad (4.13)$$

Assuming that α = 0.3 yields the effective radiative temperature of Earth as 255 K, or −18 °C. The observed temperature is, however, 288 K. This is because the radiation going to space is not emitted from the surface layers but from the colder layers at the interface of troposphere and stratosphere, and this, in turn, as we have seen above, is caused by the greenhouse gases in the atmosphere that cause the troposphere to be optically thick in the infrared. Another way of looking at this is that the surface receives an additional radiation from the troposphere of σT^4, and this will raise its surface temperature to $2^{1/4} \times 255$ K, or 303 K. While this is 15 K too warm compared to observation, it shows the physical principles at work. In reality, the atmosphere cannot be treated as a single opaque layer, and its temperature varies considerably with height, making this simple approximation, not surprisingly, come out too large. We used this elementary knowledge in Chapter 1 to make a comparison of an early greenhouse gas effect on our neighbouring planets, showing an explanation for the existence of liquid and frozen water together with water vapour in Earth's atmosphere.

We have seen that inclusion of an additional greenhouse gas in the atmosphere will tend to create a net change in the energy balance by distorting the atmospheric temperature profile (e.g. see Figure 4.6), generally resulting in a change in temperature. This change in radiation, due to any kind of perturbation—natural or anthropogenic— and defined for the tropopause level, is called radiative forcing. It is expressed in the same terms as the fluxes in Figure 4.7 (in watts per square metre) and involves some quantification of this effect over some period of time. Sometimes, however, the change at the top of the atmosphere, called the effective radiative forcing, is used. In 2013, the IPCC assessed the usefulness of several radiative-forcing metrics, distinguishing between the radiative forcing of well-mixed greenhouse gases, for which the forcing can be expressed in a globally averaged way. Examples included the key greenhouse gases carbon dioxide, nitrous oxide and methane (Intergovernmental Panel on Climate Change, 2013). Near-term climate forcers, in 'IPCC speak', are those gases that have a strong impact on climate within the first decade after their emission (the time horizon of the radiative-forcing definition); an example here is methane. Another useful concept in discussing current climate change is the Global Warming Potential. This is defined as the time-integrated radiative forcing due to a pulse emission of a given component relative to a pulse emission of an equal mass of carbon dioxide. The use of this concept is, however, confusing when dealing with variability in naturally occurring greenhouse gases, such as, for instance, methane emissions from wetlands. In that case, other metrics are more appropriate. The radiative forcing in modern times, from pre-industrial to present, for a number of components is given in Figure 4.8. To be able to calculate these numbers, global climate or Earth system models are needed (see Chapter 3). Increased carbon dioxide is the largest contributor to radiative forcing,

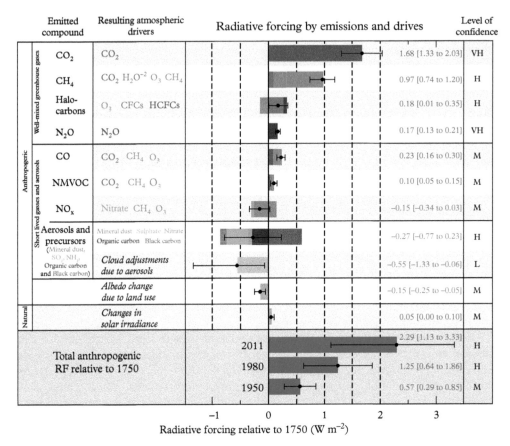

Figure 4.8 *Radiative forcing of global climate between 1750 and 2011. Horizontal bars indicate the uncertainties (95% confidence intervals). (From IPCC, 2013)*

stemming from the burning of fossil fuels; other greenhouse gases include methane, nitrous oxide and the halocarbons. A generally negative radiative forcing, indicating cooling, stems from aerosols, but the uncertainty here is very large. In fact, this is one of the key uncertainties of our current climate models (see Chapter 5). The overall radiative forcing is, however, positive and of the order of $2\,\mathrm{W\,m^{-2}}$. The large uncertainty is virtually all due to the uncertainty in the aerosols (Chapter 5). Variability in solar radiance has played a minor role in this period. It is worth emphasizing that these numbers apply to the modern era and do not include the slow forcing components such as ice sheets and oceans.

Figure 4.9 presents the current state of knowledge of the components of Earth's energy balance. Our satellite measures fluxes at the top of the atmosphere. Note that, to balance, the sum of the outgoing longwave radiation and the reflected solar radiation needs to equal the incoming solar radiation. While, at the surface, the outgoing longwave

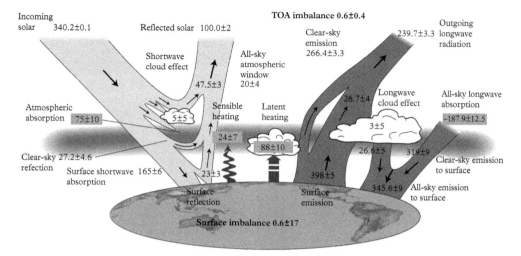

Figure 4.9 *Components of Earth's energy balance, averaged over 2000–10. Units are watts per square metre; TOA, top of Earth's atmosphere. (After Stephens et al., 2012)*

radiation is 398 W m^{-2}, at the top of the atmosphere, this is 240 W m^{-2}. This is expected, given the difference in the emitting temperatures and remembering the fact that the greenhouse effect exist because of the temperature gradient. The net longwave absorption in the atmosphere is balanced by the surface heat fluxes and the absorption of solar radiation. Note also the relatively large errors on the surface fluxes compared to the fluxes at the top of the atmosphere. The surface energy balance consists of the sum of incoming minus reflected solar, the two turbulent fluxes of sensible and latent heat, and the balance of incoming and outgoing longwave radiation. This balance, based on the currently best available observations, does not fully close and a small imbalance of 0.6 W m^{-2} remains, as at the top of the atmosphere.

5

Aerosols and Climate

5.1 Aerosols and climate

Thirty years ago, in 1987, a group of authors proposed that biota in the ocean could provide a feedback on climate other than directly through carbon dioxide (or methane). The first letters of the authors' names provided the acronym for the feedback: CLAW (Charlson, et al., 1987). The CLAW hypothesis states that phytoplankton in the ocean produce a substance called dimethylsulphide (DMS) that escapes to the air. In the air, it is oxidized and subsequently forms a sulphate aerosol that can act as a cloud condensation nucleus (CCN) for cloud formation. DMS and the sulphate aerosols are indeed abundant in the marine boundary layer. Increasing amounts of clouds can then act to reduce radiation, which would hinder further growth of the phytoplankton. Because of the sensitivity of phytoplankton growth to temperature, a potential negative climate feedback is established through the marine biosphere and aerosols that can counteract warming. The CLAW hypothesis has functioned as a very useful research hypothesis, even though it has proven hard to find evidence that substantiates the full cycle (Hatton, 2014).

On land, a similar feedback was more recently suggested through the relation between Gross Primary Production, radiation and increased levels of carbon dioxide and higher temperatures (Kulmala et al., 2004). Diffuse radiation, that is, radiation coming from more than one direction, tends to increase photosynthesis and Gross Primary Production in plants. This is something long known to agricultural producers of the greenhouses in the Netherlands; they use diffuse radiation in their greenhouses to grow their tomatoes and red peppers. The link needed to establish a feedback in the climate system is the group of volatile organic substances that are released by plants in the photosynthesis cycle. These biological volatile organic carbons (BVOCs) are released in the air, mostly as terpenes (based on the isoprene cyclopentene), where they are quickly oxidized. The terpenoid BVOCs that are then produced condense on pre-existing particles and are the largest global source of secondary organic aerosols. Secondary organic aerosols can act to reduce solar radiation, but often also act to increase the ratio of diffuse to direct radiation due to enhanced scattering of radiation (see Chapter 4). This then constitutes the positive feedback loop: more diffuse radiation enhances uptake of carbon dioxide.

Biogeochemical Cycles and Climate. Han Dolman, Oxford University Press (2019). © Han Dolman.
DOI: 10.1093/oso/9780198779308.001.0001

Secondary organic aerosols also act to reduce warming by increasing the number of CCNs; this is the negative feedback loop of the terrestrial biosphere feedback.

While both the CLAW feedback and the terrestrial feedback present an interesting hypothesis, they have been until now only supported by evidence for parts of the feedback loop and it still remains to be proven whether they operate as proposed at the global climate scale. Nevertheless, both hypotheses point to the important role of aerosols in the climate and the links with the biosphere and biogeochemical cycles. In this chapter, we further discuss aerosols and their role in climate.

Aerosols are defined in the meteorological glossary as 'a colloidal system in which the dispersed phase is composed of either solid or liquid particles, and in which the dispersion medium is some gas, usually air' (American Meteorological Society, 2019). The size definition varies but, commonly, the upper bound for the size of the particles is somewhat arbitrarily set at 10^{-5} m.

Figure 5.1 shows the variety of sizes and types of aerosols and their main sources. Aerosols are generally measured in mass units for the upper diameter of the particles. Hence, we use, for example, the values $PM_{2.5}$ and PM_{10} (in micrograms per cubic metre), for particles $\leq 2.5\ 10^{-6}$ m and $\leq 10^{-5}$ m, respectively, as indicators for the amount of aerosol of a particular category.

Aerosols, except dust, originate from a variety of chemical precursors. Sulphate aerosols resulting from anthropogenic fossil fuel burning, with sulphur dioxide as a precursor, but also; wildfires and DMS oxidation, comprise another important component. In Chapter 11, we will encounter aerosol formation from nitrogen species, primarily ammonia and nitrogen oxides. Dust resulting from erosion of dry soil brings in additional particles for aerosol formation. While, for all of these, particle size varies, the tropospheric lifetime is much more similar, and is usually around a week. Cloud scavenging and rainfall are the dominant removal mechanisms of aerosols from the troposphere, but larger particles may also just sediment.

So far, we have left one important natural source out, and that is aerosols produced by volcanic eruptions (Robock, 2007). These contain, among others, sulphur dioxide,

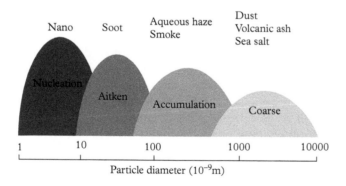

Figure 5.1 *Range of aerosol sizes and their classification. Also shown are the main sources.*

which, when further oxidized, forms sulphate aerosol. The importance of this category stems from the fact that the explosions can inject the aerosols very high into the atmosphere, well above the troposphere, into the stratosphere. Here, in the absence of the main removal mechanisms of the troposphere, they can stay active for much longer. Volcanic aerosols from large volcanic eruptions, such as those from Mount Pinatubo in 1991, have lowered Earth's surface temperature globally by almost half of a degree. The eruption of the Indonesian volcano Tambora in 1815 caused 'the year without summer' in Europe in 1816, wreaking substantial havoc on harvests (Luterbacher & Pfister, 2015). The eruption of Mount Pinatubo emitted a staggering amount of 20 million tonnes of sulphur dioxide into the stratosphere at an altitude of more than 30 km. The associated aerosols led to widespread winter warming and summer cooling and, initially, apocalyptic sunsets. The mean atmospheric temperature returned to its normal value only after two to three years. The cooling effect of stratospheric sulphate aerosols has led several scientists to propose direct injection of particles into the stratosphere as a mitigation option for climate change (see Chapters 12 and 13).

5.2 Sources and distribution of aerosols

Table 5.1 shows the contributions of natural and anthropogenic sources to global emissions of aerosols and aerosol precursors. The table shows that primary biological aerosol particles constitute a major source of natural aerosols. BVOCs, including monoterpenes and isoprene, comprise another source of natural aerosols.

Sea spray and mineral dust are also clearly important contributions to natural aerosols. However, perhaps more important to consider is the range of the estimates, as aerosol mass contributions vary substantially globally. For instance, in some continental regions with relatively high concentrations, especially in urban South East Asia and China, mineral dust dominates (>35%) the aerosol mass (Boucher et al., 2013), while over continental North and South America, organic aerosol dominates (>20%) and sea salt is dominant in oceanic environments (>50% of the mass fraction). In anthropogenic emissions, sulphur dioxide and ammonia dominate, while biomass burning is also an important contributor, not only globally but also particularly in the tropics during the burning season.

The balance of sources and sinks, as in any reservoir, determines their abundance in the atmosphere. Most aerosols have a lifetime of around a week. For dust and sea spray, sedimentation also decreases the atmospheric lifetime, sometimes down to a single day. Aerosol distributions show large variations on daily, seasonal and interannual scales as a result of considerable diversity of the sources, the short lifetime of the particles, and the dependence of sinks on the local and regional meteorology. Figure 5.2 shows an indicator for aerosol distribution, the aerosol optical depth (AOD), at 550 10^{-9} m, obtained by assimilating satellite information into a weather forecast model. The results are averaged from 2003 to 2010. Large values for AOD suggest large amounts of aerosol. The largest AODs are found over biomass-burning areas in Africa and

Table 5.1 *Global emissions of anthropogenic and natural aerosols and aerosol precursors.*
(After Boucher et al., 2013, and references therein)

Source	Min*	Max*
Anthropogenic sources†		
NMVOCs (Tg yr^{-1})	98	158
Black carbon (Tg yr^{-1})	4	6
Primary organic aerosols (Tg yr^{-1})	6	15
Sulphur dioxide (Tg S yr^{-1})	43	78
Ammonia (Tg yr^{-1})	35	59
Biomass burning (Tg yr^{-1})	29	85
Natural sources		
Sea spray (Tg yr^{-1})	1400	6800
including marine primary organic aerosols	2	20
Mineral dust (Tg yr^{-1})	1000	4000
Terrestrial primary biological aerosol particles (Tg yr^{-1})	50	1000
including spores		28
Dimethylsulphide (Tg S yr^{-1})	10	40
BVOCs		
Monoterpenes (Tg C yr^{-1})	30	120
Isoprene (Tg C yr^{-1})	410	600
SOA production from all BVOCs	20	380

Abbreviations: BVOCs, biological volatile organic carbons; NMVOCs, non-methane volatile organic SOA, secondary organic aerosol.
* The average minimum and maximum values are from a range of available inventories.
† Global anthropogenic emissions of aerosols and aerosol precursors are for 2000.

Amazonia, over desert areas and over South East Asia. Most of the aerosols reside in the lower 1–2 km of the atmosphere, except in places where outflow over the oceans occurs in the free troposphere. This is also the place where interactions between chemical species can bring serious health hazards. Elevated aerosol concentration is a known health risk and, in areas with high aerosol loading, is often an important cause of mortality among humans.

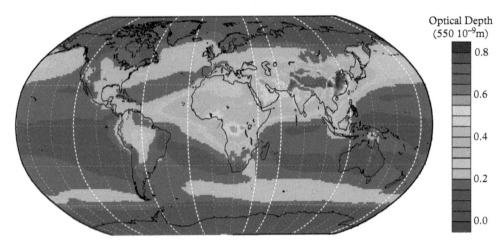

Figure 5.2 *Spatial distribution of the 550 10^{-9} m aerosol optical depth (unitless) from the ECMWF (European Centre for Medium-Range Weather Forecasts) Integrated Forecast System model with assimilation of MODIS (Moderate Resolution Imaging Spectrometer) aerosol optical depth averaged over the period 2003–10. (From Boucher et al., 2013, and further references therein)*

5.3 Aerosol–climate interaction

The interaction of aerosols with climate is complex. Aerosols interact with radiation and with clouds, with the diameter of the particles being an important factor affecting how exactly the interaction turns out. The process by which the interaction between radiation and clouds balances out into a single overall response is poorly understood. Aerosol–climate interaction can be grouped in three distinct categories: aerosol–radiation interactions, aerosol–cloud interactions and aerosol–surface interactions (Boucher, 2015). Aerosol–radiation interaction consists of scattering and absorption (see Chapter 4) of both solar and longwave radiation by aerosol particles. When solar radiation is scattered back into the sky, a cooling will be the net result. Absorption will result in a heating of the layer of air containing the aerosol but, in this case, less radiation will reach Earth's surface. Furthermore, aerosols are an important source of CCNs in liquid clouds. An increase in the availability of CCNs generally leads to more and smaller cloud droplets (with the same amount of liquid being able to condense). This leads to whiter clouds that have a higher reflectivity. This chain of events may also affect the production of precipitation. Finally, the surface interacts with aerosols, in the way we have seen for radiation and BVOCs, but also when, for instance, soot is deposited on snow or ice, thereby increasing melt and reducing the reflectivity.

5.4 Aerosol–radiation interaction

Let us now use the radiation transfer theory developed in Chapter 4 to gain some more insight into the effects of aerosols on radiation. Figure 5.3 shows the main processes involved. Aerosols scatter solar radiation in all directions; they can also absorb solar radiation.

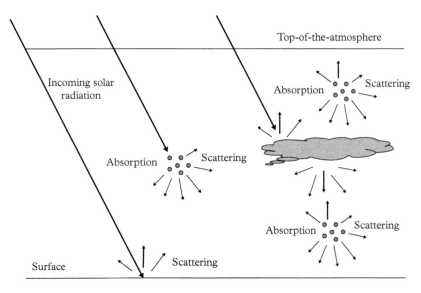

Figure 5.3 *Schematic of the aerosol–radiation interactions. (After Boucher, 2015)*

The amount and properties of solar radiation that interact with the aerosols depend on the solar zenith angle, altitude, the surface properties and the presence or absence of clouds. The reflectance R_a and its counterpart, the transmittance T_a, of a layer of aerosols in a cloud-free atmosphere and, importantly, in the absence of absorption, with an upscatter fraction β_a is given by

$$R_a \approx \frac{\beta_a \tau_a}{\cos \theta_0} \tag{5.1}$$

$$T_a = 1 - R_a = 1 - \frac{\beta_a \tau_a}{\cos \theta_0} \tag{5.2}$$

where θ_0 is the solar zenith angle, and τ_a is the optical depth. The total reflectance needs to take into account the multiple reflections from the surface (these are denoted by R_s):

$$\begin{aligned}
R_{as} &= R_a + T_a R_s T_a + T_a R_s T_a R_s T_a + T_a (R_a R_s)^2 + \dots \\
&= \frac{R_a + (1 - R_a)^2 R_s}{1 - R_s R_a}
\end{aligned} \tag{5.3}$$

which gives, after some algebra (Boucher, 2015; Charlson et al., 1991), the reflection of the fully coupled surface atmosphere system, R_{as}. The change in reflectance due to the aerosol layer is now

$$\begin{aligned}
\Delta R_{as} &= R_{as} - R_s \\
&= \frac{(1 - R_a)^2 R_s}{1 - R_s R_a} - R_s \approx R_a (1 - R_s)^2
\end{aligned} \tag{5.4}$$

Using eqn (5.1), we can calculate the amount of reflected solar radiation due to the layer of aerosol at any place as

$$\Delta F = -S \cos \theta_0 T^2 \Delta R_{as} \approx -S(1 - R_s)^2 T^2 \beta_a \tau_a \qquad (5.5)$$

with S being the solar radiation. Taking the global average (denoted by the overbar) of the relevant properties, incorporating cloud cover (A_c) and realizing that, globally, any location on Earth is only receiving solar radiation for half of the day yields the following equation for the global average change in reflection due to aerosols ΔF:

$$\overline{\Delta F} = -\frac{1}{2} S \left(1 - \overline{A_c}\right) \left(1 - \overline{R_s}\right)^2 \overline{T}^2 \overline{\beta_a} \, \overline{\tau_a} \qquad (5.6)$$

Equations 5.5 and 5.6 are simplifications (e.g. they assume no covariance between the parameters and no absorption) but yield some key insights into the radiative effects of aerosols. First, they show the interaction of aerosol properties with the surface reflectivity, or albedo, R_s. If the surface albedo R_s approaches 1, the radiative effect of aerosols goes to 0. Second, the change in radiative flux, ΔF, is linear with changes in optical depth, the upward scattering coefficient and transmittance. Third, the first three variables on the right-hand side of the equation, S, A_c and R_s, are essentially geophysical parameters relating to the physics of radiation, frequency of cloudiness and surface reflectance; the next two variables, T and β_a, determine the microphysics of the aerosols; and the last, τ_a, which is the optical depth, effectively depends on the amount and scattering properties of the species involved. This latter variable effectively provides a quantitative link to the biogeochemical cycles in which the aerosol is produced. Globally, estimates of τ_a average around 0.15. Sulphate aerosols compose about 35% of this value, and sea salt composes around 30%, the remainder being made up by smaller contributions of particulate organic matter and black carbon (Kinne et al., 2006).

Equations 5.5 and 5.6 deal with a layer of aerosol that only scatters; there is no absorption of radiation (remember, scattering does not change the wavelengths, absorption can (see Chapter 4)). This can be incorporated into the equations by writing eqns (5.1) and (5.2) as

$$R_a \approx \frac{\beta_a \varpi_a \tau_a}{\cos \theta_0}$$

$$T_a = 1 - R_a - A_a \approx 1 - \frac{\beta_a \varpi_a \tau_a}{\cos \theta_0} - \frac{\left(1 - \varpi_a\right) \tau_a}{\cos \theta_0} \qquad (5.7)$$

where now ϖ_a is the single scattering albedo of the aerosol, and A_a the absorption. Following the road that took us to eqn (5.5), we now find for aerosols with absorption and reflection that the radiative flux difference is given by

$$\Delta F \approx -S T_a^2 \beta_a \varpi_a \tau_a \left[\left(1 - R_s\right)^2 - \frac{2 R_s \left(1 - \varpi_a\right)}{\beta_a \varpi_a} \right] \qquad (5.8)$$

Comparing this equation to eqn (5.5), we can see that not only do we have an additional parameter, the single scattering albedo, but we have also lost the nice linearity of the equation.

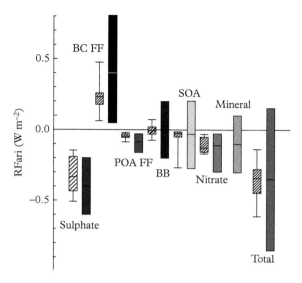

Figure 5.4 *Annual mean top-of-the-atmosphere radiative forcing due to aerosol–radiation interactions (RF$_{ari}$) due to different anthropogenic aerosol types, for the 1750–2010 period. Striped bars show the median (line), the 5%–95% uncertainty range (bar) and the min/max values (error bars) from AeroCom II models (Myhre et al., 2013) corrected for the 1750–2010 period. Solid bars show the IPCC Fifth Assessment Report best estimates and 90% uncertainty ranges; BB, biomass burning aerosols; BC FF, black carbon from fossil fuel and biofuel; POA FF, primary organic aerosol from fossil fuel and biofuel; SOA, secondary organic aerosols. (After Boucher et al., 2013)*

It requires detailed observations or radiative transfer models to show the resulting change at the top of the atmosphere. Figure 5.4 gives the current best estimates (IPCC, 2013) of the top-of-the-atmosphere radiative forcing due to aerosol–radiation feedback for the different aerosol species. Note the large uncertainty (the length of the boxes) and the uncertainty of the sign in the case of organic aerosols, for both primary and secondary aerosols. The overall radiative effect is cooling but, indeed, with a very large uncertainty.

5.5 Aerosol–cloud interaction

Aerosol–cloud interactions are shown schematically in Figure 5.5. Aerosols act as CCNs, thereby changing the natural (pre-industrial) distribution of CCNs, and thus influence the number and size of cloud droplets. It is worth emphasizing that aerosol particles also acted as CCNs in pre-industrial times; it is hard to form clouds or rain without them.

One of the ways in which aerosols can change the properties of clouds is via the so-called Twomey effect. Increasing the number of CCNs by adding aerosols will generally result in smaller cloud droplets. However, the associated increase in the total scattering cross section yields an overall increase in cloud optical density and a corresponding

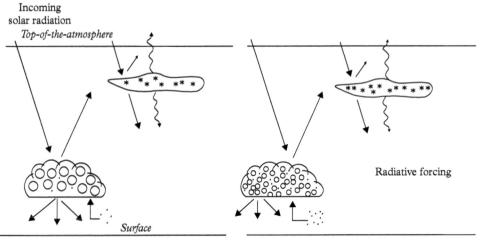

Incoming
solar radiation
Top-of-the-atmosphere

Radiative forcing

Surface

1. Unperturbed liquid water cloud and ice cloud

2. Perturbed liquid water cloud and ice cloud due to immediate changes in the number and properties of CCN and IN

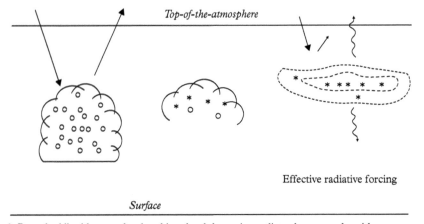

Top-of-the-atmosphere

Effective radiative forcing

Surface

3. Perturbed liquid water cloud and ice cloud due to immediate changes and rapid adjustments due to precipitation development, latent heat release due to the freezing of cloud droplet, interactions with radiation and convection, ...

Figure 5.5 *Schematic representing aerosol–cloud interactions; CCN, cloud condensation nucleus; IN, ice nucleus. (After Boucher, 2015)*

increase in reflectivity of the cloud. The cloud lines above shipping lanes provide an example of this indirect aerosol effect. Direct effects relate to the absorption and scattering of aerosol particles per se. Aerosols can also impact ice nuclei. The complex interaction in clouds between the liquid phase (and CCNs) and ice phase (and ice nuclei) makes this, next to uncertainty in the water component one of the big unknowns and sources of uncertainty in current Earth system models. Part of this complexity arises from the rapid changes or adjustments that are happening within the cloud. Condensation processes yield energy that can impact further thermodynamics, updrafts and growth of the cloud. Hence, the overall result of this myriad of changes and adjustments determines the aerosol–cloud feedback.

Figure 5.6 shows, on a log–log scale, the correlation of $CCN_{0.4}$ (the subscript refers to the value of supersaturation of the particle, below which they generally have less or no specific CCN activity; in this case, 0.4%) with AOT_{500}, the aerosol optical thickness of light at 10^{-9} m (Andreae, 2009). This graph is based on observations over a variety of sites, but the values may be different regionally and it is questionable as to whether the relation would also apply to pre-industrial conditions. Two things come out quite clearly in the diagram. First, there is a large difference in polluted versus non-polluted areas; second, in the non-polluted areas, there is hardly much difference between continental sites and maritime sites.

The effect of aerosols in clouds is important. The ability of a particle to form a cloud droplet depends on its size and composition. Cloud droplets in stratiform cloud coalesce into particles that rain out in case of the pristine clouds in non-polluted areas. In contrast, smaller drops in polluted air, as a result of increased CCN levels, do not precipitate

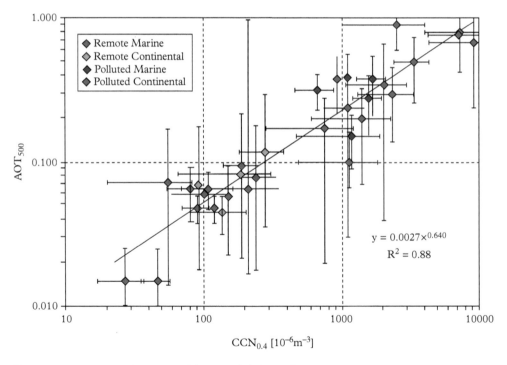

Figure 5.6 *Relationship between AOT_{500} and $CCN_{0.4}$, from investigations where these variables have been measured simultaneously or where data from nearby sites at comparable times were available. The error bars reflect the variability of measurements within each study. (After Andreae, 2009)*

before reaching the high levels in stratiform clouds, where they freeze into ice precipitation that falls and melts at lower levels. Importantly, the additional release of the latent heat of freezing at high levels, and reabsorbed heat at lower levels, by the melting ice in these polluted clouds implies a larger upward heat transport for the same amount of surface precipitation. The final outcome of this reshuffling of heat sources and sinks is a strengthening of the convective cloud and rainfall (Rosenfeld et al., 2008). These effects can be observed, for instance, by satellites by looking at cloud top temperatures. How much of a change in precipitation at the ground is due exactly to aerosol remains much more difficult to estimate, as it is hard to disentangle these effects from meteorological changes that may affect convection in a particular area.

5.6 Aerosol–surface interactions

The third and final interaction to discuss is that of aerosols with the surface. Deposition of the darker material of the aerosol, say black carbon from fires, on a lighter surface will change the albedo. When this happens on ice, the surface will be able to absorb more radiation and apply it to further melt the ice, so aerosol can have feedbacks on ice and snow melting. These changes can globally be of the order of 0.1 W m^{-2}, but are likely to vary between 0.01 W m^{-2} and 0.08 W m^{-2}.

5.7 Dust in the glacial–interglacial records

Dust particles, because of their size, are excellent tracers, and changes in the deposition of dust can be found in sediments, both marine and terrestrial (loess, lake sediment and peat record) and in ice cores. It appears from these data that the global dust flux was considerably higher during glacial stages (Maher et al., 2010). Changes in flux can record changes in source strength, because of expansion of arid and semi-arid areas or because they have become drier and more prone to erosion. This would have applied to the American Great Plains and the Russian steppes, for instance. The size of the dust particles can be used to infer changes in transport, that is, wind strength, as heavy particles cannot be transported far away from their source area. Data bases of paleo dust fluxes and sources exist that provide documentation of these changes (e.g. DIRTMAP; Kohfeld, et al., 2001).

Figure 5.7 shows changes in dust concentration from Antarctic, ocean and Greenland ice cores (Maher et al., 2010) over the last 200 kyr. The higher concentrations coincide with the glacial periods and the smaller timescale variations therein. Using isotopes, size distribution and chemical data, the origin of dust sources can be traced. This has been proven particularly successful for Chinese loess. Dust fluxes on the Antarctic during the Last Glacial Maximum have been shown to be thirty times higher than at present. The main source area is believed to have been an extended Patagonia land mass that became available for erosion as sea levels fell. Similarly, from the continental records for the

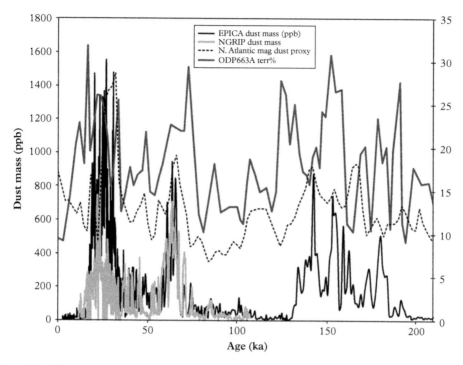

Figure 5.7 *Changes in glacial–interglacial dust concentrations over a pole-to-pole transect. (After Maher et al., 2010. Data from north Northern Atlantic, Antarctica (EPICA), Greenland (NGRIP) and equatorial Atlantic (ODP))*

United States, Eurasia and China, fluxes during the Last Glacial Maximum have been estimated to have been considerably larger.

So, overall, the picture is that there was greater dust deposition during the Last Glacial Maximum and during glacial times. But what was the cause? Was it a change in transport, a change in source area or a change in dryness? In China loess, increases in particle size have been interpreted as changes in the strength of the monsoonal (winter) winds. As an example, the size of quartz particles at a location in the China Central loess plateau ranges from 6^{-9} 10^{-9} m for the Holocene to 15 10^{-6} m in the final stages of the glacial period (Maher et al., 2010). Increased fluxes and larger particles may indicate both larger source areas and increased winds that are better able to transport the larger particles.

The high dust deposition in the ocean is also thought to have played an important, but as yet controversial, role in the enhancement of export productivity in the ocean (see also Chapters 9 and 12). The impact on the production in the equatorial Pacific appears to be small while, in the Southern Ocean, such enhancement is thought to have contributed to the lower values of atmospheric carbon dioxide by about 40–50 ppm. This observation shows the important role of aerosols in climate regulation, one that may help to explain the so-called faint sun paradox (see Chapter 10). Not only do they provide feed-backs to the climate system through radiation and cloud interactions but, through their content and chemical composition, they can interact directly with biogeochemical cycles. The debate about the link between acid rain and sulphate aerosols provides a strong reminder of these impacts. When sulphate aerosols were phased out through legislation, the cleaner air lost some of its capacity to cool the atmosphere, and the effects of green-house-gas-induced climate change became much more visible after the 1980s (see also Chapter 12).

6

Physics and Dynamics of the Atmosphere

6.1 The atmosphere as a heat engine

The concept of the atmosphere developed in this chapter is that of a complex thermo-dynamic engine. This engine obeys the classical laws of physics in a three-dimensional context. In principle, we can describe the mean state of the atmosphere with the following variables: pressure, temperature, humidity and wind speed and direction. In a thermo-dynamic engine, energy is transferred from one particular point in space to another, involving some amount of mechanical work. In Earth's atmosphere, heat is transferred from the surface to higher levels, involving complicated three-dimensional atmospheric motions. The energy supplying heat for this transport is the radiation from the Sun, as characterized in Chapter 4. This overall picture of a thermodynamic engine is further extended by phase changes, that is, condensation and evaporation play an additional role in shaping the patterns and motions of the atmosphere.

Figure 6.1 shows an atmospheric interpretation of a thermodynamic engine, with air rising in the tropics and descending near the poles, creating a poleward heat transport. During these processes, heat is converted into kinetic energy, driving the motion. We can calculate the efficiency of the atmospheric heat engines by referring to the Carnot theorem. This states that the efficiency, or the maximum amount of work as a fraction of the heat input, that can be derived from such an engine is

$$\eta = \frac{T_h - T_c}{T_h} \tag{6.1}$$

with T_h and T_c being the temperatures of the hot and cold reservoirs, respectively. For our example of poleward transport of heat from the tropics, with a tropical temperature of, say, 300 K and a polar one of 250 K, the efficiency of the transport would be about 17%. For vertical transport, we take the surface temperature of the planet at 288 K and the temperature at the level at which longwave radiation (see Chapter 4) is emitted back to the atmosphere as 255 K. This yields an efficiency of 11%. In reality, both numbers are considerable lower, of the order of a few per cent, but these figures provide an upper limit from thermodynamic theory and provide a general framework for interpretation.

Biogeochemical Cycles and Climate. Han Dolman, Oxford University Press (2019). © Han Dolman.
DOI: 10.1093/oso/9780198779308.001.0001

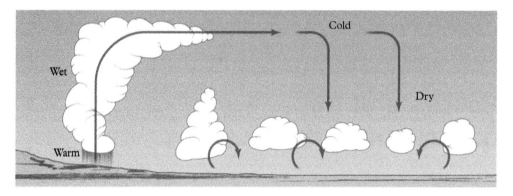

Figure 6.1 *Schematic of an atmospheric heat and moisture engine in which heat from a hot, wet reservoir at the tropics is transported to the cold and dry poles, involving motions and phase changes. (After Pauluis, 2015)*

6.2 Basic atmospheric thermodynamics

Energy is conserved following the first law of thermodynamics and thus the amount of work done by the heat engine, dW, must equal the difference between the flux of supplied heat dQ and the change of internal energy dU. More formally,

$$dU = dQ - dW \tag{6.2}$$

For heat exchange in the atmosphere, we introduce the concept of a parcel of air, a virtual set of molecules that moves quickly enough so that exchange of energy with its surroundings can be neglected and pressure and volume changes can occur almost instantaneously. The internal energy of such a parcel of air depends only on the temperature change dT and is given as

$$dU = c_v dT \tag{6.3}$$

where c_v is the specific heat at constant volume. In general, the amount of work performed equals force times the distance but, in the atmosphere, it is easier to work with pressure (which is force per unit area), so we get

$$dW = pdV \tag{6.4}$$

where p is the pressure, and dV is the change in volume (distance times area). We can now write the first law of thermodynamics for an air parcel as

$$dQ = c_v dT + pdV \tag{6.5a}$$

or, per mass unit,

$$dq = c_v dT + pd\alpha \tag{6.5b}$$

where α is the specific volume (in cubic metres per kilogram), and dq and c_v are per mass unit. (The specific volume α is the reciprocal of the mass density, $1/\rho$.) Equation (6.5b) is sometimes easier to use, as it effectively deals with a unit mass of a substance.

The ideal gas law specifies the exact relation between pressure, volume and temperature:

$$p\alpha = RT \tag{6.6a}$$

or, in differential form,

$$pd\alpha + \alpha dp = RdT \tag{6.6b}$$

with R being the gas constant per kilogram of dry air. Combining eqn (6.6b) with eqn (6.5b) yields

$$dq = (c_v + R)dT - \alpha dp \tag{6.7}$$

We learn from eqn (6.7) that, if the pressure is kept constant, increasing the temperature by dT requires a heat input $dq = (c_v + R)\, dT$; in other words, the specific heat at constant pressure, c_p, is

$$c_p = c_v + R \tag{6.8}$$

6.3   The tropospheric lapse rate and potential temperature

We are now in a position to derive one of the key characteristics of the lower atmosphere, the gradient of temperature with height, also called the tropospheric lapse rate that played an important role in the derivation of the greenhouse effect in chapter 4. Let us assume a parcel of air with a temperature T and a specific volume α at pressure p. If the parcel rises adiabatically (no exchange of heat between its environment and the parcel; i.e. $dq = 0$), using eqns (6.7) and (6.8), the first law of thermodynamics can be written as

$$c_p dT = \alpha dp \tag{6.9}$$

We can combine this with the hydrostatic equation, which gives us the relation between the pressure of the atmosphere, and the mass and gravitation. The hydrostatic equation states that, for air in hydrostatic equilibrium, the weight of the atmosphere between levels z and $z + dz$ (the right-hand side of the equation) is balanced by the difference in pressure between the levels dp, such that

$$dp = -\rho g dz \tag{6.10}$$

where g is the acceleration due to gravity, 9.81 m s^{-2}. Eliminating dp from eqns (6.9) and (6.10) and using $\alpha = 1/\rho$ gives us a relation for the change in temperature with height:

$$\frac{dT}{dz} = -\frac{g}{c_p} = -\Gamma_d \qquad (6.11)$$

The specific heat for air at constant pressure is 1010 J K^{-1} kg^{-1}, which yields a value of $\Gamma_d = 9.7$ K km^{-1}. Anyone who has ever tried to climb a mountain in a Mediterranean climate, such as Mount Ventoux in southern France in summer, will have experienced the very large differences in temperature between the foot of the mountain and the top. They are largely due to eqn (6.11). But, let us also emphasize the assumptions we have made to derive the lapse rate. First, the atmosphere is in hydrostatic equilibrium; in practice, this means that we do not deal with small-scale motions but rather work with averages over horizontal scales larger than, say, 10 km. Second, the atmosphere is without moisture; including moisture, as we will see in Section 6.4, complicates the picture substantially and reduces the gradient. Third, there is no heat input into the parcel, the adiabatic constraint. And, fourth, mixing between the parcel and its environment plays no role. In the real atmosphere, these conditions are easily violated.

Using the gas law (eqn (6.6a)) and the first law of thermodynamics (eqn (6.5)), it is possible to derive a very useful quantity in atmospheric research, the potential temperature:

$$\theta = T\left(\frac{p_0}{p_1}\right)^{R/c_p} \qquad (6.12)$$

where p_1 is the pressure at the current level, and p_0 is the pressure at a reference level (usually 10^5 Pa at the surface). Poisson's law states that the potential temperature is a conserved quantity, that is, under dry adiabatic constraints, the potential temperature does not change (e.g. Wallace and Hobbs, 2006). For instance, when air moves over a mountain, it cools as it expands and then warms back up when it descends; the potential temperature, unlike the actual temperature, does not change during this process if the parcel does not saturate with moisture or gets additional energy input.

6.4   Moisture in the atmosphere

Moisture in the atmosphere changes many of the fairly simple relations for the dry adiabatic atmosphere discussed previously. The moisture-mixing ratio w is defined as

$$w = \frac{m_v}{m_d} \qquad (6.13)$$

with m_v and m_d being the mass of water vapour and the mass of dry air, respectively. The specific humidity is defined as the mass of water vapour over the total mass of air in a unit volume:

$$q = \frac{m_v}{m_v + m_d} = \frac{w}{1+w} \qquad (6.14)$$

For practical purposes, q (not to be confused with the heat per mass unit) closely follows the value of w, since w is generally $\ll 1$.

When moisture condenses, an amount of heat is released, L_v. The latent heat of vaporization is an important parameter in meteorology as its value is very large, indicating that, in the process of evaporation and condensation, large amounts of energy can be exchanged. The value of L_v decreases slowly with temperature but, at 0 °C, it is 2.5×10^6 J kg^{-1}.

The water vapor has a (partial) pressure e which is related to q in the usual approximation by

$$e = \rho q R T / 0.622 \tag{6.15a}$$

(ρq is the water mass per volume unit, and 0.622 is the molar mass ratio of water to dry air). Since $p = \rho R T$, it follows that

$$e / p = q / 0.622 \tag{6.15b}$$

Water vapour in the atmosphere has a peculiar property—the amount of water vapour the atmosphere can hold is finite. Specifically, there is a saturation vapour pressure $e_s(T)$ above which the water vapour will condense (see also chapter 1). This can be illustrated by considering a closed box with a layer of water at the bottom. Molecules of water from the water layer will evaporate into the air in the box, and a number will condense back from the air into the water. If the amount of molecules evaporating equals the amount condensing, the air is said to be 'saturated' with respect to water vapour. If fewer molecules are condensing than are evaporating, the air is said to be 'unsaturated'. The saturation vapour pressure changes approximately exponentially with temperature, following the Clausius–Clapeyron equation:

$$\frac{de_s}{dT} = \frac{L_v e_s}{RT^2} \tag{6.16}$$

where L_v is the latent heat of vaporization, and e_s is the saturated vapour pressure at temperature T. A practical approximation for e_s is given by the equation

$$e_s(T) = 6.1094\, e^{\left(\frac{17.625 T_c}{T_c + 243.04}\right)} \tag{6.17}$$

with T_C in degrees Celsius, and e_s in hectopascal (hPa). This equation tells us that, for each degree of temperature increase, the atmosphere can hold 6%–7% more water. Thus, with air temperature increasing as a result of climate change, we expect more water vapour in the atmosphere and an intensification of the hydrological cycle (see Chapter 8).

When an unsaturated parcel of air rises, expands and cools down, it may reach the saturation vapour pressure level for that temperature. Although no external heat is supplied, the heat released at condensation then provides heat to the parcel, making the process moist adiabatic. The first law of thermodynamics, with no heat added from outside now has an additional term:

$$c_v dT = -p d\alpha - L_v dq \tag{6.18}$$

with the last term signifying the heating realized by condensation (the condensed mass per kilogram of dry air equals $-dq$). This allows us to calculate the moist adiabatic lapse rate, the change of temperature with height when condensation occurs. Proceeding as we did earlier in deriving eqn (6.11), we can rewrite this in the form

$$c_p dT = -gdz - L_v dq \tag{6.19}$$

Using $dq = d(0.622\, e_s(T)/p)$, the gas law (eqn (6.6b)), the hydrostatic equation (eqn (6.10)), the Clausius–Clapeyron equation (eqn (6.16)) and $c_p = c_v + R$, it can be shown that

$$\Gamma_{sat} = -\frac{dT}{dz} = \frac{\Gamma_{dry}\left[1+(\varepsilon L_v e_s / pRT)\right]}{1+(\varepsilon^2 L_v^2 e_s / c_p pRT^2)} \tag{6.20}$$

with the factor $\varepsilon = 0.622$ (e.g. Taylor, 2005). This shows that the effect of condensation in the atmosphere is to reduce the temperature gradient of a wet atmosphere; this reduction can be up to 4 °C km^{-1}.

Figure 6.2 shows the dry and saturated (wet) adiabatic curves for a range of temperatures in a so-called skew-T diagram. In cold environments (shown on the left-hand side of the graph), the saturated and dry adiabats are very similar because e_s is small; at high

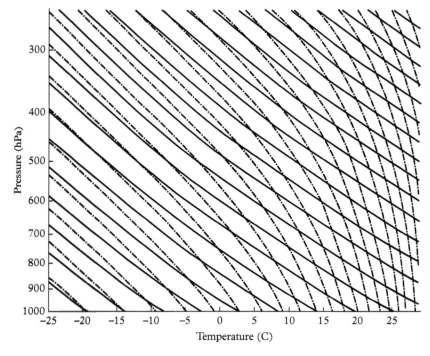

Figure 6.2 *Schematic skew-T diagram, showing the wet (dashed lines) and dry (solid lines) adiabats as a function of temperature and pressure.*

temperatures, the saturated adiabat is considerably steeper. Since moisture in the atmosphere is mostly confined to the lower troposphere (high T and $e_s(T)$), the moist and dry adiabats are almost equal at high altitude (low T and $e_s(T)$). The temperature dependence makes the difference larger in the tropics and smaller near the poles. For instance, if we take an air parcel at 20 °C, it will follow the saturated adiabat when saturated, and the dry adiabat when unsaturated. The temperature then, at 200 hPa, will be −55 °C and −80 °C for the saturated and unsaturated parcels, respectively. Now, in the real atmosphere, rising air parcels cool until they become saturated, so parcels will often follow the unsaturated adiabat first and then, when they become saturated, the saturated adiabat. The point where parcels become saturated is called the Lifting Condensation Level and it plays an important role in understanding the formation of clouds. A slightly different version of the diagram shown in Figure 6.2, also using moisture lines (isohumes), can be used to plot radiosonde data to analyse the thermodynamics of the atmosphere. This allows to predict the Lifting Condensation Level, the amount of growth of a cloud into a thunderstorm, for instance, and the amount of precipitable water. Such a complete thermodynamic diagram also predicts the effects of föhn winds, which occur when a parcel first cools for some time following a dry adiabat but then rains out following the wet adiabat and, downwind of a mountain, produces a very dry and hot environment through descending a dry adiabat of the same potential temperature (remember, this is a conserved property).

6.5   The equations of motion

Let us now further sketch the properties of the three-dimensional atmosphere by looking more closely at the forces acting on a parcel of air. To do that, we start by introducing the vector notation. Up to now, we have mainly been concerned with scalar properties. But force, acceleration and velocity not only have a magnitude but also, and importantly, a direction. Vector relations have the nice property that they remain invariant in a coordinate system. We can use them in a Cartesian coordinate frame in the well-known x, y and z directions, as well as in a system with spherical coordinates, latitude, longitude and distance to Earth's centre, as this is often more useful for analysis of the properties of the atmosphere on a rotating sphere, such as Earth. As atmospherics dynamics is fundamentally a three-dimensional problem, vector notation provides an efficient way of approaching the equations that govern motion in the atmosphere. Vectors are shown in bold font.

When a parcel of air with mass m experiences a force $\mathbf{F}$, it will, according to Newton's second law, experience an acceleration $d\mathbf{V}/dt$ in the direction of that force:

$$\frac{d\mathbf{V}}{dt} = \frac{\mathrm{F}}{m} \tag{6.21}$$

Here, $\mathbf{V}$ is the vector with component directions u, v and w. The forces acting on a parcel of air in the atmosphere are the pressure gradient force, arising from differences in pressure; a frictional force, which depends on velocity gradients but is usually roughly counter to the direction of the velocity; the gravitational and centrifugal force; and a sideward force, the Coriolis force. The centrifugal and the Coriolis force are 'apparent' forces arising from Earth's rotation. We can then write the acceleration of **V** as the sum of these forces:

$$\frac{d\mathbf{V}}{dt} = \frac{1}{m}\Sigma \mathbf{F} \qquad (6.22)$$

Starting with the pressure gradient force, this can be expressed as

$$\frac{\mathbf{F}}{m} = -\frac{1}{\rho}\nabla p \qquad (6.23)$$

where ρ is the density of the air, and p is pressure; this equation indicates that winds are forced from high pressure to low pressure along the pressure gradient. The direction of this wind is initially at right angles to the isobars (lines of equal pressure). The closer the isobars are together, the steeper the pressure gradient, and the stronger the wind will be. However, as soon as the wind starts to blow, it becomes subject to the Coriolis force that deflects its path.

The Coriolis force appears because Earth rotates: Earth moves with an angular velocity of Ω 7.27 × 10^{-5} radians s^{-1}. We can derive the magnitude of the Coriolis force as follows (Holton & Hakim, 2012). Assume a mass of air that is moving zonally with a velocity u at latitude ϕ but which is now displaced and moves southward (see Figure 6.3). Conservation of momentum requires that the angular momentum at the old and the new position remain the same. This requires the angular velocity of the parcel to change by adjusting its zonal velocity by δu. What is the size of this change? Equating the momentum at the two positions over a distance $R + \delta R$ with δu, the change in displacement velocity, gives

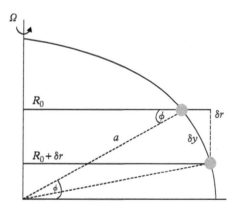

Figure 6.3 *Geometry to calculate the magnitude of the Coriolis force. In this example, an air mass with horizontal speed u is suddenly displaced southwards (equatorial) from a latitude ϕ towards the equator over a distance δy, R_0 is the distance to the north–south rotation axis at the original, and $R + \delta R$ is the new, larger distance.*

$$\left(\Omega + \frac{u}{R}\right) R^2 = \left(\Omega + \frac{u + \delta u}{R + \delta R}\right)(R + \delta R)^2 \qquad (6.24a)$$

Solving for the change in velocity neglecting second order terms in δR and δu yields

$$\delta u = -2\Omega\delta R - \frac{u}{R}\delta R \qquad (6.24b)$$

Since $R = a \cos\phi$, where a is radius of the Earth and ϕ the latitude (Figure 6.3), dividing eqn (6.24b) by δt and then letting δt going to zero (Holton, 2004) yields the equation for the zonal component of the Coriolis acceleration:

$$\frac{du}{dt} = 2\Omega v \sin\phi + \frac{uv}{a}\tan\phi. \qquad (6.25)$$

Similar equations can be derived for the wind components in the other directions. When the parcel continues to move, it will accelerate now westward towards an observer on Earth. This is the acceleration caused by the Coriolis force, $\boldsymbol{F}_C$. The first term of the right-hand side of eqn (6.25) is the 'Coriolis term'; the second is referred to as the 'metric term' or 'curvature effect' and is generally two orders of magnitude smaller than the first for typical synoptic systems. It is often neglected for practical purposes. The Coriolis force acts perpendicular to the motion and thus performs no work. Since the Coriolis term is positive for the Northern Hemisphere and negative for the Southern Hemisphere, the Coriolis force in the Northern Hemisphere is directed to the right of wind, while it is directed to the left of the wind in the Southern Hemisphere. At the equator, there is no Coriolis force. The Coriolis force is of great importance in describing the larger scale dynamics of the atmosphere and ocean.

The Coriolis acceleration can be expressed in vector form for a moving parcel of air with velocity $\mathbf{V}$ as

$$\frac{d\mathbf{V}}{dt} = -f\mathbf{k}\times\mathbf{V} \qquad (6.26)$$

where $\mathbf{k}$ is the unit vector, and f is the Coriolis parameter ($f = 2\Omega\sin\phi$). In this equation, the cross product $-\mathbf{k}\times\mathbf{V}$ indicates a rotation of $\mathbf{V}$ by 90° clockwise with respect to the velocity. In Figure 6.4, a sketch is given of the Coriolis forces that an air parcel experiences when it gets displaced eastward, with the directions of the Coriolis force given by the small arrows.

We have discussed the pressure gradient force and the Coriolis force and need to add, for completion of the three-dimensional picture, the gravitational acceleration, as $\mathbf{g}$ becomes an important part of the force field in the vertical. We assume for simplicity that the centrifugal force is part of the term. The general equation of motion in vector form can then be expressed as

$$\frac{d\mathbf{V}}{dt} = -2\Omega\times\mathbf{V} - \frac{1}{\rho}\nabla p + \mathbf{g} + \mathbf{F}_\mathbf{r} \qquad (6.27)$$

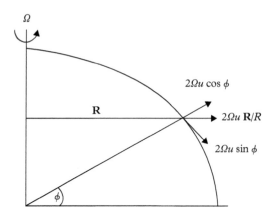

Figure 6.4 *Sketch of components of the Coriolis force that work along a vector **R** to deflect an air parcel that moves in the eastward direction along a latitude circle.*

where **V** is velocity. Here, the vector Ω should be expressed in the local coordinate system: $\Omega = \Omega(0, \cos\phi, \sin\phi)$. In eqn (6.27), the first term on the right-hand side is the acceleration from the rotating Earth, and the second and third terms represent the pressure gradient acceleration and the gravitational acceleration, respectively, while $\mathbf{F}_r$ is the frictional force (or acceleration per unit mass) that works against the direction of the velocity. This equation forms the basis of much of dynamical meteorology.

Expanding eqn (6.27) in the three Cartesian component directions (x, y, z) and neglecting the terms dealing with curvature (e.g. eqn (6.25)) yields

$$\frac{du}{dt} = -\frac{1}{\rho}\frac{\partial p}{\partial x} + 2\Omega v \sin\phi - 2\Omega w \cos\phi + F_x \tag{6.28a}$$

$$\frac{dv}{dt} = -\frac{1}{\rho}\frac{\partial p}{\partial y} - 2\Omega u \sin\phi + F_y \tag{6.28b}$$

$$\frac{dw}{dt} = -\frac{1}{\rho}\frac{\partial p}{\partial z} + 2\Omega u \cos\phi - g + F_z \tag{6.28c}$$

The friction terms in these equations are here given per unit mass and so are equivalent to acceleration. Furthermore, the term containing the cosine in eqn (6.28a) is often neglected, as it is much smaller than the sine term.

While these equations may look rather formidable at first sight, we can derive some very useful insights from them. Let us start by assuming that, in the free atmosphere (the higher part of the atmosphere, where the effects of friction are no longer felt), the frictional force can be ignored ($F_i = 0$). The horizontal flow in eqns (6.28 a, b) is then only dependent on the Coriolis force and the pressure gradient force. We can derive an expression that gives us the wind velocity as the balance between the pressure gradient force and the Coriolis force:

$$fv = \frac{1}{\rho}\frac{\partial p}{\partial x} \tag{6.29a}$$

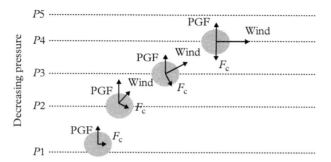

Figure 6.5 *Geostrophic flow with a north–south pressure gradient from p_1 to p_5. The parcel at p_1 begins to accelerate under the pressure gradient force (PGF). It then experiences a Coriolis force perpendicular to the direction of the velocity. The parcel continues to accelerate until, at p_4, the Coriolis force is equivalent but opposed in sign to the PGF, and the wind velocity is parallel to the isobars. This is geostrophic balance.*

$$fu = -\frac{1}{\rho}\frac{\partial p}{\partial y} \tag{6.29b}$$

The balance between the pressure force and Coriolis force is called the geostrophic balance, and the wind is denoted by $\mathbf{V}_g$. In vector form, this is

$$\mathbf{V}_g = \frac{1}{\rho f}\mathbf{k}\times\nabla p \tag{6.30}$$

Winds in geostrophic balance approximate the real winds in mid-latitude systems within 10%–15% of the true horizontal velocity, so it is indeed a very useful approximation. Wind in geostrophic balance flows parallel to the isobars, as is shown in Figure 6.5.

6.6   The thermal wind equation

It is one thing to derive wind velocities from pressure gradients, but where do these gradients arise from and how can we calculate winds from them? The gradients in pressure primarily arise from differences in temperature that translate into density differences and hence into horizontal pressure gradients. Air pressure decreases more rapidly with height in cold environments than in warm, less dense air. This causes the isobaric planes (planes of constant pressure) to tilt upward into the direction of the warm air (see Figure 6.7).

Now, using the hydrostatic equation (eqn (6.10)) and the gas law (eqn (6.6a)), we can show that

$$Z_2 - Z_1 = -\frac{R\overline{T}}{g}\ln(p_2/p_1) \tag{6.31}$$

where $\bar{T}$ is the pressure-weighted mean of T. This equation is called the hypsometric equation and shows that the thickness of a layer bounded by two isobaric surfaces with a gradient in height is related to the mean temperature of that layer.

We can also write the geostrophic wind velocity in isobaric coordinates; in this case, we have using eqs. 6.29 and 6.10 and the triple product rule

$$fu_g = -g(\partial z / \partial y)_p \tag{6.32a}$$

$$fv_g = g(\partial z / \partial x)_p \tag{6.32b}$$

The subscript 'p' in the right-hand term indicates that z is here considered to be a function of x and y within the isobaric surface, with p constant (the definition of an isobaric plane). These equations tell us that the geostrophic wind speed is proportional to the slope of the isobaric surface and is directed 90° to the right/left of the downslope direction for the Northern/Southern Hemisphere.

Taking it now one step further, we will derive an equation that gives the velocity change with height as a function of the horizontal temperature gradient only. The resulting equation is the so-called thermal wind equation and is derived using the gas law equation (eqn (6.6b)) and the hydrostatic equation (eqn (6.10)) by differentiating eqn (6.32) to p (again, the subscript p means that p is kept constant on partial differentiation) (e.g. see Holton & Hakim, 2012). Implicit in the derivation of the thermal wind equation is that the thickness of a layer bounded by two isobaric surfaces is proportional to the mean temperature of that layer (eqn (6.31)):

$$\frac{\partial u_g}{\partial \ln p} = \frac{R}{f}\left(\frac{\partial T}{\partial y}\right)_p \tag{6.33a}$$

$$\frac{\partial v_g}{\partial \ln p} = -\frac{R}{f}\left(\frac{\partial T}{\partial x}\right)_p \tag{6.33b}$$

In vector form, the thermal wind equation reads:

$$\frac{\partial \mathbf{V}_g}{\partial \ln p} = -\frac{R}{f}\mathbf{k} \times \nabla_p T \tag{6.34}$$

where T is temperature. The thermal wind equation tells us something very important: it tells us that the rate of change of geostrophic wind with height change (also called shear) is dependent on the horizontal temperature gradient. Note that it is not a real wind: it is the vector difference of the geostrophic wind at two different levels:

$$\mathbf{V}_t = \mathbf{V}_g(p_2) - \mathbf{V}_g(p_1) \tag{6.35}$$

where $p_2 < p_1$. This is shown graphically in Figure 6.6, where the thermal wind vector is shown as the difference between the two geostrophic wind vectors at pressure p_1 for the lower level, and p_2 for the upper level ($p_1 > p_2$).

The thermal wind equation was originally derived for the atmosphere but, importantly, it can also be derived for the ocean, where the geostrophic flow depends then on the horizontal density gradient (see Chapter 7, eqns (7.3a, b)).

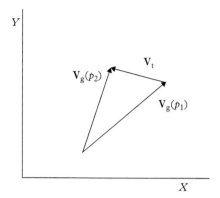

Figure 6.6 *Sketch of the thermal wind vector* $\mathbf{v}_t$ *as the difference between the geostrophic wind vectors at* p_1 *and* p_2.

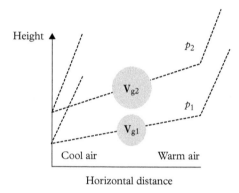

Figure 6.7 *Cross section of the atmosphere with cool air and low pressure and a sharper gradient in pressure with height on the left side, and warm air at high pressure and a lower gradient on the right. The geostrophic wind at the lower level is smaller than that at the upper level. The circles give the magnitude of the resulting geostrophic wind (directed into the page) following from eqn (6.32), with the larger circle having the larger magnitude.*

Let us illustrate this important relation with some examples. In a barotropic atmosphere (an atmosphere in which density depends only upon pressure), the temperature is constant on the constant pressure surfaces, and no vertical gradient in wind speed exists. In the more realistic situation, where the temperature surfaces intersect the pressure surfaces and do not run parallel, the wind changes with height, following eqn (6.29) (see Figure 6.7). This situation is called baroclinic. Knowing the three-dimensional temperature distribution and pressure at the surface, it is now possible to fully define the geostrophic wind field. For instance, for a gradient of temperature of 0.75 K per degree of latitude at around 30° latitude, we obtain a wind speed of around 37 m s⁻¹ at 250 hPa (note that we assume there is zero geostrophic wind at the surface).

When steep temperature gradients exist, for instance, at the boundary between polar air masses and mid-latitude air, the thermal wind equation can be used to predict the strength of the jet stream, a strong westerly wind at around 9 km height.

Winds can turn supergeostrophic when wind speed is accelerated along the direction of the wind. To achieve this, the Coriolis force needs to be larger than the pressure gradient force. This can happen at high latitudes. Supergeostrophic winds can occur at the entrance of the core of a jet stream. Conversely, subgeostrophic winds occur when the Coriolis force is less than the pressure gradient force. In this case, lateral acceleration takes place perpendicular to the wind direction. These situations are associated with cyclonic and anticyclonic flow, phenomena most often associated with mid-latitude weather. A further complication arises when friction plays an important role, say, in the lower parts of the troposphere. In this case, the reduction of wind speed and corresponding change in the Coriolis force's direction causes low-pressure systems around the land, where friction is high, to fill up (convergence) and high-pressure systems to empty (divergence).

6.7   Weather systems and global climate

The unequal distribution of radiation over Earth's surface is the principal reason for large differences in temperature from the tropics to the poles, and why the excess heat is transported from the tropics to the poles. At long timescales, comparable to the ice ages, the distribution of incoming solar radiation over the planet and seasons is determined by Milankovitch cycles (see Figure 6.8). Milankovitch was a Serbian mathematician and astronomer who meticulously calculated Earth's orbit from the changing positions of Earth and the Moon and the laws of gravitation.

The 'eccentricity' determines the deviation from roundness of Earth's orbit around the Sun. It increases and decreases with a period of 100,000 years. Earth's north–south axis is tilted with respect to the plane of orbit and changes with a period of 41,000 years; this is called the 'obliquity cycle', and the 'wobble' of this axis changes the time of the year when Earth is closest to the Sun, with a period of 19,000–23,000 years. This cycle is called the precession cycle. These three cycles together play a role in the pace of the

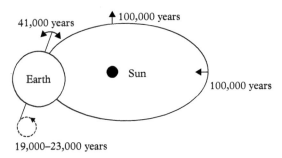

Figure 6.8 *Earth's Milankovitch cycles, caused by eccentricity (100,000 years), obliquity (41,000 years) and precession (19,000–23,000 years).*

ice ages in the last 1.5 million years. In a now-famous paper, Hays et al. (1976) concluded that 'changes in the earth's orbital geometry are the fundamental cause of the succession of Quaternary ice ages'; importantly, they also stated that a 'model of future climate based on the observed orbital–climate relationships, but ignoring anthropogenic effects, predicts that the long-term trend over the next several thousand years is towards extensive Northern Hemisphere glaciation'. Crucially, of course, this prediction of the next ice age neglected anthropogenic influences on the climate due to emissions of carbon dioxide from the burning of fossil fuel. With the availability of ice cores from Antarctica, covering several glacial cycles (Chapter 2), this theory was largely confirmed, although there are still a number of teething problems. The changes in insolation that are important are particularly those above 65° N, where decreased solar radiation in summer may allow snow to remain throughout the year and start to form ice sheets.

Let us move towards current day conditions and the present-day solar input. Figure 6.9 shows the latitudinal distribution of incoming solar radiation as a function of time of year. The radiation at the top of the atmosphere varies by about 3.5% over the year, as Earth moves around the Sun. As we have seen, this is primarily because Earth's orbit is elliptical. Earth is thus closer to the Sun at one time of year (the perihelion, early January) than at the opposite time (aphelion, early July). The time of year when Earth is at perihelion varies with the precession period of 21,000 years. The tropics show

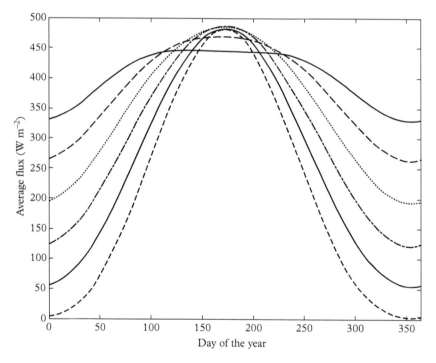

Figure 6.9 *The daily average solar irradiation on a plane horizontal to Earth's surface for different latitudes ranging from 5° N to 85° N (solid line at the top to the lowest dashed line).*

considerably less variation in solar radiation than the higher latitudes. While the strength of solar radiation of the Sun is almost the same for all latitudes, the day length and solar zenith angle (the angle between an imaginary point above a particular location and the centre of the Sun's disc) are not, and this causes the main differences in latitudinal radiation.

As shown in Chapter 4, Earth reflects back part of the solar radiation to space and absorbs and converts the remainder to longwave radiation that leaves Earth. The net balance between the absorbed solar radiation and the outgoing longwave radiation determines which part of Earth has an energy deficit or surplus. Figure 6.10 shows how the imbalance between absorbed solar and outgoing longwave radiation gives rise to an energy surplus in the tropics, and energy deficits in mid-latitudes and polar regions. The heat transport associated with this energy gradient is from the equator to the poles.

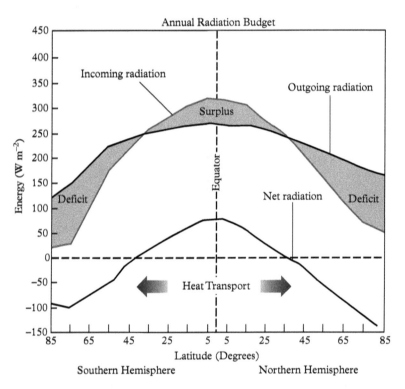

Figure 6.10 *The balance of absorbed incoming shortwave radiation and outgoing longwave radiation and net radiation as a function of latitude, showing the surplus at the tropics and the deficit towards the poles. The direction of the poleward-associated heat transport is shown by the thick arrows. (The source of this graph is the COMET® Website at http://meted.ucar.edu/of the University Corporation for Atmospheric Research (UCAR), sponsored in part through cooperative agreement(s) with the National Oceanic and Atmospheric Administration (NOAA), U.S. Department of Commerce (DOC). ©1997–2017 University Corporation for Atmospheric Research. All Rights Reserved.)*

The strong gradient in incoming radiation between the poles and the equator is the main cause of the variable weather found at mid-latitudes; it drives baroclinic instabilities that form extratropical cyclones. Because of the rotation of Earth, the transport of heat through the atmosphere is broken up into three cells per hemisphere (Figure 6.11(a)). So, for the Northern Hemisphere, and mirrored in the Southern Hemisphere, the first zone from the equator to roughly 30° is called the Hadley cell, the transport around mid-latitudes (30°–60° N) is called the Ferrel cell, and the cell around the poles is called, simply, the Polar cell. Near the equator, moist air rises and forms large convective towers that continue to grow by the release of latent heat from condensing water vapour (see Chapter 8, Figures 8.2 and 8.8). Because of the lack of horizontal pressure gradient, the horizontal winds are weak. Hence, sailors in the days of sailing ships called this region the 'doldrums'. When the rising air is constrained by the tropopause (the border between the troposphere and the stratosphere), the air is spread out poleward. As the air moving poleward cools by emitting longwave radiation, it becomes denser and forms high-pressure areas at latitudes of around 30° N and 30° S. This converging air produces the sub-tropical highs. When the air slowly descends because of its density, it warms by compression (in contrast to the ascending air that expands and cools). Over the areas where we find subtropical highs, we also find the world's main desert areas, which receive virtually no rain.

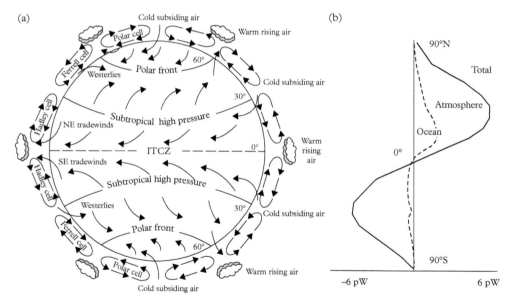

Figure 6.11 *Schematic diagram of wind and pressure distribution over Earth, and associated heat transport. (a) Diagram of main global wind patterns and pressure distribution. The Hadley, Ferrel and Polar cells are also shown; ITCZ, Intertropical Convergence Zone. (After Stuart Chapin et al., 2011) (b) Distribution of meridional heat transport by ocean (dashed line), atmosphere (difference between total and ocean) and total heat transport (solid line), based on a model, for the last 22,000 years. (After Yang et al., 2015)*

The surface air moving back towards the equator also becomes subject to the Coriolis force and forms the trade winds that gave the old-time sailors reliable north-easterly winds in the Northern Hemisphere. The latitude band around 30° is also called the horse latitude, the etymology of this term remaining somewhat unclear. When the north-easterly trade winds collide with the south-easterly trade winds, the air rises strongly and produces the Intertropical Convergence Zone (ITCZ). This is an area of high precipitation. The ITCZ shifts seasonally with the shift in the maximum incoming radiation and appears as a band of clouds, usually thunderstorms, that circle the globe near the equator; this indicates the ascending branch of the Hadley cell.

In the Ferrel cell, the northward-bound transport gives rise to westerlies as the winds are deflected eastward by the Coriolis force. This westerly flow is, unlike the trade winds, not constant. Sequences of low and high-pressure cells interchange as they move from west to east. Indeed, a defining feature of the flow of this region is the importance of cold and warm fronts that form the transition zone between air masses of different densities, caused by temperature differences such as those found along the north-eastern seaboard of North America. In cold fronts, a wedge of cold air pushes forward and upward a layer of warm air. Cold fronts are characterized by sharp temperature gradients over short distances, sharp differences in the humidity content of the air and strong shifts in wind direction, with associated differences in air pressure. They originally form as part of a baroclinic disturbance. They are also characterized by their type of clouds (cumuliform, often bringing showers) and the ratio of horizontal to vertical length scales is of the order of 50:1. The front itself is often characterized by a narrow band of heavy thunderstorms that give the front its typical gusty wind and heavy rainfall. In warm fronts, warm air is pushed forward and upward over a retreating wedge of cold air. Warm fronts have a horizontal-to-vertical length ratio of 300:1, are much broader and are typically characterized by more layered clouds such as stratus and nimbostratus (Ahrens & Henson, 2016).

Big whirls in the horizontal wind field are called cyclones if they rotate in Earth's direction (anticlockwise/clockwise for Northern/Southern Hemisphere). They occur when the surface wind is converging to a centre and deflected of its course by the Coriolis force. The converging air then replaces the ascending air in the centre, which is usually associated with formation of clouds and precipitation. Big whirls rotating opposite to Earth are called anticyclones. They occur when the surface wind is divergent, in which case the Coriolis force causes an opposite sense of rotation. This goes together with descending air, in which clouds tend to evaporate.

Fronts can further develop into extratropical cyclones (note that tropical cyclones have a different origin: warm seawater rather than a polar front is the direct source of energy) and the model that is most often used to describe this development from an initial baroclinic wave into a full blown extratropical cyclone is that of the Bergen School. Conditions favourable for the genesis of these cyclones are a strong baroclinity, weak atmospheric stability, a large moisture input and a large amplitude in the jet stream, allowing horizontal divergence aloft and the formation of strong updrafts. These updrafts are associated with convergence at the surface, and the deflection of this motion by the Coriolis force causes the typical cyclonic rotation. Cyclogenesis is constrained to the mid- and high (polar) latitudes. This is due to the role of vorticity (vorticity is a vector quantity and expresses the tendency of a fluid particle to rotate or circulate at a particular point; hence, vorticity increases towards the poles (see also Chapter 7, eqn (7.4)). A typical lifetime of a cyclone is about a week. Extratropical cyclones are the main mode of atmospheric heat transport in the mid and high latitudes (see Figure 6.11(b)) and can cause severe damage in the form of wind damage, and floods. Extremely fast developing cyclones, characterized by a >24 hPa drop in the central pressure within 24 hours, are called 'bomb' or 'explosive' cyclones. They can cause major havoc when they reach land, through their strong winds and associated extreme precipitation (see Figure 6.12). One of the crucial research questions is how the frequency and intensity of such storms may change with future climate.

Tropical cyclones or hurricanes form in the tropical ocean when the sea-surface temperature is higher than 26 °C. Unlike their extratropical counterparts, which tend to have a cold core and originate from temperature differences in the horizontal plane, tropical cyclones have a warm core with little horizontal temperature variation. They are associated with extremely strong winds and large amounts of rainfall. The energy for the formation of tropical cyclones comes primarily from the release of latent heat when

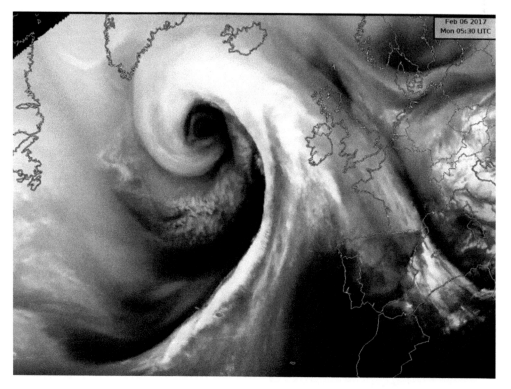

Figure 6.12 *Bomb or explosive cyclone in development on 2 February 2017 over the North Atlantic, as seen by infrared satellite. The pressure at the centre was dropping below 740 hPa. Note the classic cold core and the comma-like development of the associated cloud. (Copyright 2017 EUMETSAT)*

condensation occurs (see Figure 6.2). This forces the air to rise further and creates a massive low-pressure centre in the middle of the cyclone; around this, a ring of clouds forms. The low-pressure centre draws in moist air, sustaining the thermodynamic engine of the cyclone. When tropical cyclones make landfall, they cause considerable damage but, because the moisture flow into the centre breaks down, they quickly lose their strength. Over the Atlantic, tropical cyclones originate as small disturbances over land that travel westward on the easterlies and, when conditions are suitable, may develop into a full tropical cyclone. Like the extratropical cyclones, their occurrence and frequency under climate change is a major research question, with some scientists suggesting there may be more tropical cyclones as a result of climate change, but others suggesting there may be fewer or the same number as at present but that they will be of higher intensity.

7

Physics and Dynamics of the Oceans

7.1 Earth's oceans

The oceans cover 70.8% of Earth's surface, a non-negligible part. They comprise three main basins: the Pacific Ocean (181.34×10^6 km^2), the Atlantic Ocean (107.57×10^6 km^2) and the Indian Ocean (74.12×10^6 km^2). Their average depth is 3.7 km, with most of the ocean depths ranging between 2.0 and 6.0 km. In contrast, the land has a mean elevation of 1.1 km, while the atmosphere extends an order of magnitude more. A typical profile of the ocean floor, with exaggerated vertical versus horizontal scaling, is given in Figure 7.1. Where the continental land meets the ocean at the continental margins, shelf waters exist that function essentially as overflow basins of the ocean at large timescales; typical examples are the North China Sea and the North Sea. At the mid-ocean ridges, new crust is continuously being created through the tectonic cycle (see Chapter 1). Trenches, such as the Mariana Trench, which is east of the Philippines in the Pacific, are known to be up to 11,034 m deep. Oceanic trenches originate where one tectonic plate subducts beneath another one (Stewart, 2008).

Fundamentally, the ocean behaves like the atmosphere, as a fluid, and is subject to the same constraints of fluid dynamics as the air on a rotating sphere. This implies that some of the equations we used in Chapter 6, such as the equations of motion and the thermal wind equation, also hold for the ocean. There are, however, some important differences. The oceans have a much higher heat capacity than the atmosphere: in fact, the total heat capacity of the global atmosphere could be contained in just the upper 3.2 m of the oceans. This is because the specific heat of water is four times that of air, and of course, the oceans have much higher density: the density of water is about 1000 times that of the air at sea level. In the atmosphere, radiation plays an important role in generating temperature and pressure differences. In the ocean, this is far less important, as only the first few metres of the ocean absorb virtually all radiation. The ocean, however is actively involved in the heat budget of the climate system—the oceans are a great moderating influence on climate change. At long timescales (>1000 years), the ocean is crucial to understanding changes in our climate. Since 1850, Earth's ocean has absorbed most (>95%) of the excess energy produced by the enhanced greenhouse effect.

Biogeochemical Cycles and Climate. Han Dolman, Oxford University Press (2019). © Han Dolman.
DOI: 10.1093/oso/9780198779308.001.0001

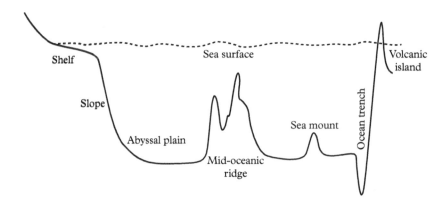

Figure 7.1 *Schematic diagram of the ocean floor with a mid-oceanic ridge, a trench and a sea mount vertical scale exaggerated. (After Stewart, 2008)*

7.2   Density, salinity and temperature

Let us first take a look at one of the key features of the ocean, its salinity. Ocean water contains about 3.5% salt, most of which (85%) is sodium chloride. It also contains smaller quantities of magnesium, sulphur, calcium and potassium. Salinity originates from weathering and riverine input. However, it is important to appreciate that removal in the ocean at geological timescales plays a major role in the salinity budget. These removals may occur by a variety of geological and biological processes. For instance, calcium is a dominant element in river water but, in ocean water, it is far less abundant. The removal of calcium from ocean water is achieved by, among other unicellular organisms, foraminifera, which produce calcareous skeletons which contribute to carbonate sediments, playing an important role in the long-term carbon cycle (see Chapter 9).

Salinity in the open ocean results from the balance between evaporation of water (increasing salinity) and precipitation (decreasing salinity) and can be observed using either in situ measurements from ships or buoys, or measurements from recently developed remote sensing techniques. Figure 7.2 gives the distribution of salinity for 1955 to 2012, derived from in situ observations. Note first the small variation in surface salinity over the globe, from 3.3% to 3.7%. This makes the measurement of salinity gradients difficult. The North and South Atlantic basins show the highest salinity. This is due to the fact that evaporation exceeds the input through precipitation and river flow. Salinity decreases towards the poles, where precipitation and river input exceed evaporation. Meltwater from sea ice and ice sheets as a result of increasing air temperature has decreased salinity further in the northern latitudes. So, despite the massive input of salt and sediments from rivers in the Atlantic, such as the Amazon, the overall salinity is very much a result of only the balance between evaporation and precipitation. The ocean transport from the Atlantic to other basins keeps the Atlantic from becoming too salty.

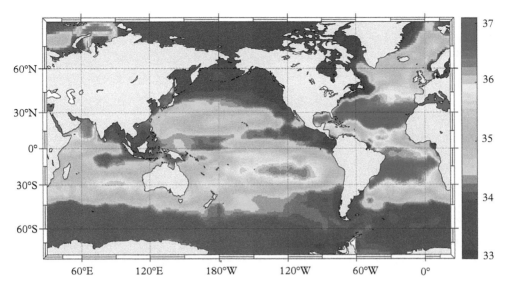

Figure 7.2 *Sea-surface salinity from in situ measurements. (Data from https://www.metoffice.gov.uk/ hadobs/en4/) (Good, Lowe, Collins, & Moufouma-Okia, 2008). Graphic courtesy of Meike Sena Martins, University Hamburg.)*

The mean depth-averaged temperature of the ocean is 3.5 °C, with 50% of the variability in the range 1.3–3.8 °C (Stewart, 2008). This temperature is remarkably constant. The sea-surface temperature, however, is considerably more variable. Around the equator, the sea-surface temperatures can go up to 25–30 °C, while near the poles, of course, the temperatures drop to close to zero or just below (Figure 7.3). Note also the rather homogeneous zonal temperature in the Southern Hemisphere, while that of the Northern Hemisphere is perturbed by the large continental land masses. Sea-surface temperature is a climatological variable of tremendous importance; it plays a large role in the generation of cyclones (see Chapter 6), the generation of rainfall (see Chapter 8), and the fundamental ocean and ocean–land oscillations such as the Atlantic Meridional Oscillation and the El Niño–La Niña cycle (see Section 8.5).

Salinity, temperature and pressure determine the density of sea water. The relation between these three variables is given by the so-called equation of state, which requires measurements at different salinities, temperatures and pressures, and subsequent curve fitting. This is in stark contrast to the atmosphere, where we have the equation of state that is derived from first principles, the gas law (Chapter 6). Nevertheless, useful approximations have been developed that can work at the required resolution to calculate the density differences that are needed to estimate flow rates. Figures 7.2 and 7.3 show examples of the horizontal distribution of surface salinity and temperature over the ocean.

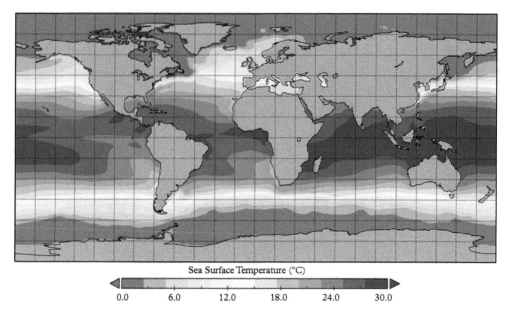

Sea Surface Temperature (°C)

0.0   6.0   12.0   18.0   24.0   30.0

Figure 7.3 *Long-term mean sea-surface temperatures from 1955 to 1964, from the 2013 World Ocean Atlas. (Accessed at https://www.nodc.noaa.gov/OC5/woa13fv2/on 18 May 2018.)*

It is now time to complete the three-dimensional picture of the ocean by looking at the vertical distribution of temperature and salinity. There is only a shallow layer of the ocean that exchanges heat and momentum with the atmosphere; 98% of the deep ocean is not directly coupled to this exchange layer and here the temperature and salinity remain nearly constant with depth. Importantly, this shallow exchange layer or mixed layer varies with depth through the year with season. This is shown in Figure 7.4 (Kara, 2003). The mixed layer is separated from the top layer by a so-called thermocline, in which temperature declines rapidly with depth to the more stable values in the deep ocean. The depth at which the thermocline varies seasonally is called the seasonal thermocline. The permanent thermocline extends from 1500 m to 2000 m.

In the mixed layer, the wind generally produces sufficient turbulence for the layer to be well mixed. This results, in general, in a shallow mixed layer in summer, and a deep one in winter, in each of Earth's hemispheres. Shallow summer mixed layers appear with summer heating of the upper ocean together with relatively weak winds. Deep-winter mixed layers are generated by winter cooling and strong winds. In Antarctic regions, the mixed layer is shallow throughout the year. In the tropics, the variability over the year is generally small (Kara, 2003), while the Atlantic shows the largest variability throughout the season in mid-latitudes (Figure 7.4).

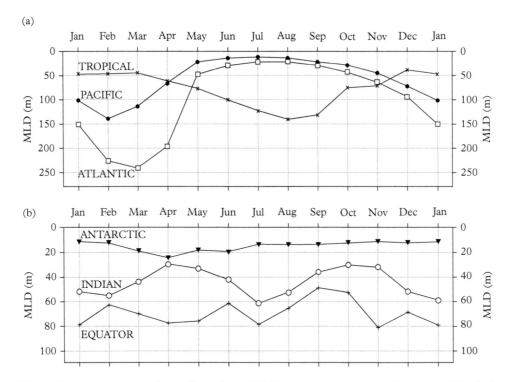

Figure 7.4 *Annual changes of mixed layer depth (MLD) at six locations in the global ocean: (a) Tropical Ocean (20° S, 140°W), Pacific Ocean (30° N, 160° E) and Atlantic Ocean (40° N, 30°W); (b) Antarctic (70° S, 100°W), Indian Ocean (10° N, 5° E) and the equator (1° S, 170° E). (After Kara, 2003)*

The salinity is nearly constant in the mixed layer and, at places where evaporation exceeds precipitation, say, between 10° and 40° latitude, the mixed layer is saltier than the thermocline. Towards higher latitudes, the freshwater input through rain and melting ice is higher than evaporation, and the mixed layer may be less salty than the thermocline.

7.3   Ekman flow

The gravitational and apparent forces working on a parcel of water in the ocean have large scales compared with most others, such as the mechanical forces arising from surface wind stress, internal pressure gradient, atmospheric pressure variations and seismic sea floor motions (think of tsunamis). Fundamentally, these forces are similar to those acting in the atmosphere (Chapter 6). Let us first look at the effect of a steady wind on the ocean surface. This is the work of Fridtjof Nansen and Walfrid Ekman. Nansen, the well-known explorer, was one of the first to notice that the wind in Arctic regions tends to blow icebergs and, in fact, the ship he was sailing on, the *Fram*, at angles of 20°–40° to the mean wind direction. Note that this provides a first indication that Earth's rotation may play an important role. If there were no impact of Earth's rotation, then the icebergs' direction of motion would, in principle, be the same as the direction of the wind, with friction causing the flow rate to decrease with depth. However, with rotation, the three important forces are the wind stress W, the friction force F_r and the Coriolis force F_c (see Chapter 6 for a discussion of the Coriolis force). Nansen realized that, for steady flow, these forces would have to balance: $W + Fr + Fc = 0$. Ekman then derived the equations for this motion. Figure 7.5 gives a schematic sketch of these forces operating on an iceberg. The frictional force always operates counter to the main direction of the iceberg's motion, while the Coriolis force, as we have seen in Chapter 6, acts orthogonally to the motion. Concerning the flow, at the surface of the ocean, the Coriolis force effectively deflects the waters to the right of the wind in the Northern Hemisphere.

Ekman showed that, deeper, in what is now known as the Ekman layer, increasing drag forces a change of direction of the flow according to

$$u = V_0 e^{az} \cos(\pi / 4 + az) \tag{7.1a}$$

$$v = V_0 e^{az} \sin(\pi / 4 + az) \tag{7.1b}$$

where V_0 is the flow at the surface, z is the depth (counted as negative) and a is a constant that depends on the eddy diffusivity and the Coriolis coefficient. These equations hold for wind stress directed towards the north (or with the y-axis aligned with the wind stress) and make use of the momentum balance equations (Chapter 6). A similar analysis can be done at the bottom of the ocean for the benthic layer, invoking the so-called no-slip condition at the seabed where the flow necessarily becomes zero. These surface equations show two important phenomena: (i) flow decreases exponentially with depth towards the ocean bottom (or with height, in the case of the benthic layer) and

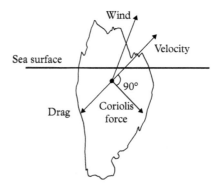

Figure 7.5 *Forces acting on an iceberg under a steady wind on a rotating Earth. The direction of the iceberg is not in line with the wind direction as a result of the balance of the Coriolis force, the drag and wind working on the iceberg.*

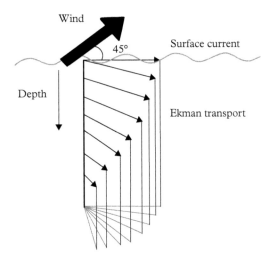

Figure 7.6 *Schematic diagram of the Ekman layer for the Northern Hemisphere, showing decrease of the flow with depth and orientation away from the windflow, leading to a net mass transport of 90° to the wind direction at the surface. The surface flow is oriented 45° to the direction of the wind.*

(ii) associated with that change in velocity is a change in direction (Figure 7.6). Below the Ekman layer, where friction can be ignored, the flow approaches geostrophic flow. The net mass transport by the Ekman flow consists of the integrals of eqns (7.1a) and (7.1b) and is oriented 90° to the surface wind direction (Figure 7.6). When the mass transport is divided by the density, the volume flow is obtained. In ocean science, the volume flow integrated over a cross section is usually expressed as million cubic metres per second, a unit called the sverdrup. The depth of the surface Ekman layer in the ocean varies with the wind speed at the surface and latitude; typical depths range from 40 m to 300 m. The lower depth values occur with low wind speeds at higher latitudes.

The analysis of the Ekman layer and transport assumes a steady wind speed, which exerts a constant stress at the interface of the ocean and air. When the wind stress is variable, so-called inertial motions may develop. These are very common in the ocean. Imagine a parcel of water suddenly being brought into motion by a gust of wind. The parcel starts moving but, once the wind drops, no other forces, except friction working against the velocity, are working on the parcel, so the Coriolis force drives the direction of the motion. This results in anticyclonic and cyclonic flow patterns in the Northern and Southern Hemispheres, respectively. Of course, at some time, the flow become part of other currents or simply decays.

Ekman flow in the Northern hemisphere becomes particularly important near north–south-oriented coastlines to the west of the continents, such as those of California, Peru and Morocco. As we have seen, surface flow with an equatorward wind direction is towards the west. This will move surface waters away from the western coasts (Figure 7.7). At the bottom of the Ekman layer, the reverse holds and water is transported towards the coast. Putting it simply, this water then replaces the waters moved away by the surface

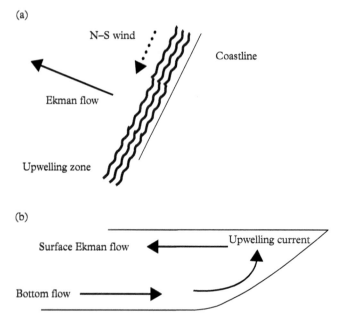

Figure 7.7 *Schematic diagram of the effect of Ekman flow near a north–south-oriented west coast with a north–south wind. Panel (a) shows a planar view; Panel (b) shows a cross section.*

Ekman flow. In general, this deeper water is colder and contains more nutrients; hence, we find some of the world's largest fisheries in upwelling regions.

7.4   Geostrophic flow in the ocean

Similar to the fluid motion in the atmosphere, we have geostrophic flow in the ocean (Apel, 1988). Using the momentum balance equations (see Chapter 6) and noting that pressure variations in the vertical depend only on the mass of the column, we can derive the geostrophic equations for flow in the ocean as

$$\frac{1}{\rho}\frac{\partial p}{\partial x} = fv \tag{7.2a}$$

$$\frac{1}{\rho}\frac{\partial p}{\partial y} = -fu \tag{7.2b}$$

$$\frac{1}{\rho}\frac{\partial p}{\partial z} = -g \tag{7.2c}$$

Note the similarity with the atmospheric equations discussed in Chapter 6. Remember also that these equations are derived for frictionless flow only; they basically describe the flow of the ocean below the Ekman layer. They state that, in the Northern Hemisphere, the flow of a south-to-north steady current, v, is deflected eastwards by the Coriolis force and is balanced by the westward horizontal pressure gradient force. Similarly, for a west-to-east flow, u, subject to a southward-directed Coriolis force, the flow will be balanced by a northward pressure gradient force. These forces are typically operating at large scales, both spatially (>50–100 km) and temporally (over several days).

Like wind, ocean flows obey the thermal wind equation (chapter 6). In this case, the pressure gradient depends on the depth below a reference level and, in parallel to the atmospheric Buys Ballot rule, the flow would have colder, denser water appearing to the left-looking downstream in the Northern Hemisphere. The equations read as follows:

$$-f\frac{\partial v}{\partial z} = g\frac{\partial}{\partial x}\left|\ln\left(\frac{\rho}{\rho_{\text{ref}}}\right)\right|_p \tag{7.3a}$$

$$f\frac{\partial u}{\partial z} = g\frac{\partial}{\partial y}\left|\ln\left(\frac{\rho}{\rho_{\text{ref}}}\right)\right|_p \tag{7.3b}$$

The subscript 'p' again indicates that p should be kept constant when differentiating, not z. These equations show the gradient of velocity with depth, that is, the shear, not the velocity itself (as in the atmosphere), as a function of horizontal density variations. When we compare them to the atmospheric equations, we note the similarity, but we have now explicit reference to density, which in the ocean relates to temperature and salinity. In the ocean, we thus need hydrographic observations of salinity and temperature both in the horizontal and in the vertical (depth) to be able to calculate flow rates.

We can now make a picture of geostrophic flow for the ocean, as we did for the atmosphere. This is shown in Figure 7.8. It is important to realize that, in the ocean, the topography of the ocean bottom influences the topography of the ocean surface far more than the currents do. For this reason, the height difference in Figure 7.8 is often expressed as the geopotential $\Phi = gh$, where h is the height over which a mass is moved perpendicular to the geoid surface (remember, in the atmosphere, we can replace to a good approximation the geopotential height with geometric height). Barotropic flow in the ocean is similar to that in the atmosphere when the height surfaces do not intersect the density surfaces; baroclinic flow arises when they do intersect.

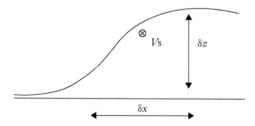

Figure 7.8 *Slope of the pressure surface used to calculate geostrophic flow in the ocean. The geostrophic current V_s is into the page and given by eqn (7.2a). Note that, in the ocean, the ratio $\delta z/\delta x$ is of the order of $1:10^5$, so the x-axis and the y-axis are not to scale.*

7.5 The ocean circulation

We have so far only described the variability of the surface-layer current with depth. Let us now expand this view into describing open ocean circulation features. One of these features is a large rotating system called a gyre; gyres rotate clockwise in the Northern Hemisphere, and anticlockwise in the Southern Hemisphere. The direction of the circulation of these gyres can be easily understood from the prevailing directions of the wind fields which drive them: east to west at the equator through the trade winds, and west to east at mid-latitudes (see Figure 6.11a). The circulation pattern in these gyres is, however, strongly asymmetric, which is less easy to understand. For instance, in the North Atlantic, the flow northward along the east coast of the United States, known by its more familiar name, the Gulf Stream, is much more restricted in area than the equatorward flow that dominates the gyre. Similar patterns occur in the Pacific, with a strong intensification off the Japanese coast. Figure 7.9 demonstrates this more clearly by showing the July 2016 average global meridional flow obtained from a state-of-the-art ocean data assimilation system (Balmaseda et al., 2012). The closer we get to the west boundary of a gyre, the more the meridional transport component increases. This is known as western intensification. In the Northern Hemisphere, the flow towards the equator occupies a much larger area than the poleward return flows at the western boundary. This is largely in line with the theory of vorticity or spin.

Vorticity is a measure of the rotation of an object. Positive vorticity is in the direction of Earth's rotation and produces anticlockwise spin; negative vorticity produces the reverse, clockwise spin. Mathematically vorticity, ζ, is expressed as the vertical component of the curl of the velocity vector:

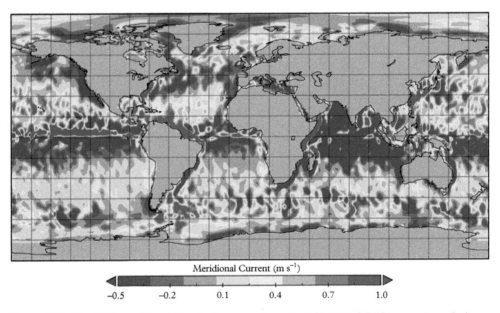

Meridional Current (m s^{-1})

-0.5 -0.2 0.1 0.4 0.7 1.0

Figure 7.9 *July 2016 meridional flow in the ocean. From the ECMWF ORAS4 ocean reanalysis product. (Balmaseda et al., 2012)*

$$\zeta = \nabla \times \mathbf{v} = \frac{\partial v}{\partial x} - \frac{\partial u}{\partial y}. \tag{7.4}$$

This vorticity is relative to an observer who is moving with Earth as it rotates. The full or absolute vorticity consists of two components: the planetary vorticity and the relative vorticity. Planetary vorticity depends on Earth's rotation and equals the Coriolis parameter, which declines with decreasing latitude (Chapter 6). The other component is the relative vorticity, the vorticity that is a result of the flow itself in the ocean. Now, the absolute vorticity, ζ_a, is given by the sum of these two:

$$\zeta_a = \zeta_p + \zeta_r \tag{7.5}$$

where the subscripts 'p' and 'r' indicate planetary vorticity and relative vorticity, respectively. In frictionless flow, which, as we assume, occurs in the deep parts of the ocean, vorticity is conserved, so any negative change in the planetary vorticity with latitude will have to be balanced by an increase in relative vorticity of the flow.

To be able to explain the area dominance of the equatorward flow (see Figure 7.9), we proceed as follows (e.g. Stewart, 2006). Neglecting horizontal pressure gradients, balance of forces requires that the surface stress τ be balanced by the Coriolis force:

$$\tau_x = -\rho h f v \tag{7.6a}$$

$$\tau_y = \rho h f u \tag{7.6b}$$

where h is the depth of the oceanic boundary layer, and u and v are the mean velocities within this layer. Taking the curl yields

$$\frac{\partial \tau_y}{\partial x} - \frac{\partial \tau_x}{\partial y} = f[\frac{\partial}{\partial x}(\rho h u) + \frac{\partial}{\partial y}(\rho h v)] + \rho h v \frac{df}{dy} \tag{7.7}$$

Because the mean winds have a mainly zonal direction, $\partial \tau_y/\partial x$ can be neglected compared to $\partial \tau_x/\partial y$ when averages are taken; the expression within square brackets is zero because it expresses the mass divergence, which has to be zero in the long term (mass balance); and df/dy, the variation of the Coriolis parameter with latitude, is called β. So, we end up with

$$v = -\frac{1}{\rho h \beta} \frac{\partial \tau_x}{\partial y} \tag{7.8}$$

When these conditions are met, the system is said to be in Sverdrup balance. This equation predicts an overall equatorward flow. This cannot overall be true, since mass balance requires a return flow (provided by generated surface slopes and consequent horizontal pressure forces which we have neglected), but the argument helps to understand why the return flow is reduced to a surprisingly limited area in the western boundary currents. These current systems play a very important role in the heat transport from the equator towards the poles.

Below the surface layer lies the deep ocean, the abyss where water flows in the abyssal, or deep, circulation. The coupled surface and deep ocean flow is often referred to as the thermohaline circulation. The deep ocean circulation is particularly important for climate. With the circulation, vast amounts of heat are moved around the globe at timescales of decades to millennia. These flows influence climate. We have also seen in Chapter 1 how important the deep ocean at longer timescales is for the storage of anthropogenic carbon dioxide. The deep ocean circulation involves much more mass than the surface flow; indeed, the deep ocean comprises more than 90% of the ocean, with just 10% of the ocean, or a few hundred metres, comprising the surface layer. The difference in temperature determines the stratification of the ocean, with cold water (around 4 °C) below the thermocline, and warmer water (up to 25 °C and more) at the surface. We saw in Figure 6.11(b) that the oceans carry almost half of the energy absorbed in the tropics to higher latitudes. The importance of this transport for climate expresses itself at various timescales. The heat transport system can be thought of as a global conveyor belt, a term originally coined by Wallace Broecker in 1987. In the Northern Hemisphere, the warm surface currents in this belt move excess energy northwards and lose energy due to cold winds and evaporation at the higher latitudes. There are only a few places where deep water is formed through this sinking motion: near the Greenland Sea and the Labrador Sea in the north, and near the Weddel Sea and the Ross seas in the Antarctic (see Figure 7.10). In addition, in the Antarctic, a process called brine enrichment takes place where freezing of water pushes out the salt of the ice, thus making the remaining water heavier and causing it to sink. From these deep-water formation places, the cold and salty water moves towards the tropics; these water masses are known as the North Atlantic Deep Water and the Antarctic Bottom Water, respectively. The deep water comes back to

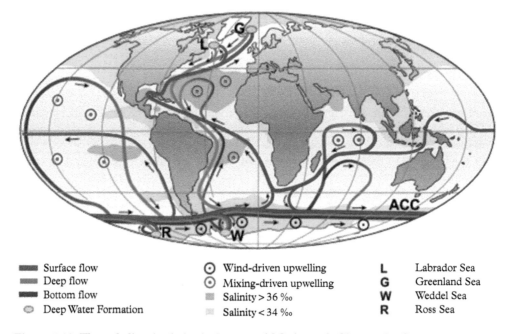

Figure 7.10 *Thermohaline circulation in the ocean; ACC, Antarctic Circumpolar Current. (After Rahmstorff, 2006)*

the surface in the equatorial regions, through a process called 'upwelling'. This is a mixing process that is strongest near mid-ocean ridges, sea mounts and boundary currents (see Figure 7.1). A word of caution: the thermohaline circulation as pictured in Figure 7.10 is a good and useful abstraction. However, in reality, the flow is much less organized and is composed of eddies of various sizes and whose precise flow is influenced by ocean floor topography, surface wind variations and continental boundaries.

7.6   Ocean and climate

An important final note about the role of ocean flows and associated heat transport. When the balance between salt and fresh water changes as a result of, for instance, increased runoff of rivers into the Arctic Ocean, or enhanced ice sheet melting, the deep-water formation may be affected. In a classic paper, Stommel (1961) used a simple box model with an equator to pole temperature difference that included salinity. Salinity provides a positive feedback loop in that simple system when higher salinity in the deep-water formation area enhances the circulation. The circulation then transports (advects) higher salinity waters into the deep-water formation regions. In this model, the high-latitude salinity increases linearly with the flow, and the flow increases linearly with high-latitude salinity. Combining these two linear equations in salinity yields a quadratic equation for the flow as function of freshwater input. This equation leads to two possible equilibrium states (e.g. as described in Chapter 1). If freshwater input to the northern Atlantic is systematically increased and

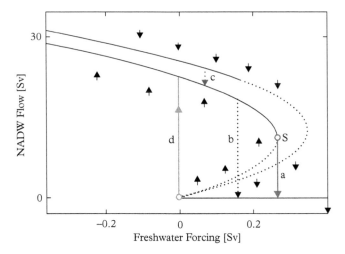

Figure 7.11 *A so-called stability diagram for the Atlantic thermohaline circulation. Solid lines indicate stable equilibrium states, and dotted lines indicate unstable states. Possible transitions are indicated by arrows: (a) reduction in salt advection, (b) shutdown in deep-water formation, (c) transition between different patterns of deep-water formation and (d) restart of deep-water formation. 'S' delineates the bifurcation beyond which no deep-water formation is possible; NADW, North Atlantic Deep Water. (After Rahmstorf, 2006)*

decreased, a hysteresis loop forms between a transition where salt advection is reduced (labelled 'a' in Figure 7.11) and a transition where deep-water formation is restarted (labelled 'd' in Figure 7.11).

This hysteresis behaviour also shows up in state-of-the-art climate models. In the Quaternary Period, three major circulation modes can be identified from the sediment record: (i) a warm mode similar to that in present-day Atlantic, (ii) a cold mode with deep water forming south of Iceland and (iii) a total breakdown mode without significant formation of North Atlantic Deep Water (Figure 7.11).

In this simple model, the temperature difference drives the circulation, whereas the salinity breaks it. Note that this model is essentially based on equator-to-pole temperature and salinity differences. Models based on pole-to-pole differences, incorporating both hemispheres, generated more complex behaviour, but this conceptual model gives a good indication of the sensitivity of climate to ocean circulation.

The El Niño–Southern Oscillation is the other famous ocean–climate interaction. This oscillation causes big changes in the dynamics of the equatorial regions (see also Chapter 8). It starts when trade winds in the western Pacific weaken and the normally deep thermocline becomes shallower. This causes a stagnation of the east-to-west flow allowing the heat to accumulate off the west coast of South America. This then deepens the thermocline in the East Pacific. The normally westward-located warm pool subsequently moves towards the central Pacific and, with it, the area of strong convection and rainfall. As a result, Indonesia and Australia experience drought and soaring high temperatures. In contrast, along the south coasts of Brazil and west South America, increased rainfall produces massive flooding. El Niño also impacts other areas of the world, and generally the high peaks in the global temperature record (see Chapter 2) are associated with El Niño events.

8

The Hydrological Cycle and Climate

8.1 The global water cycle

The importance of water on Earth was made clear by the iconic imagery of a blue planet seen by the astronauts of the first manned spacecraft. Indeed, we do live on a blue planet with 71% of Earth's surface covered by ocean and ever-changing clouds transporting water through the atmosphere. It is water that makes the planet habitable because of its greenhouse gas absorption. It is also water that plays a key role in food production. Where there is no water, food production becomes almost impossible. Finally, it is water that transports large amounts of heat away from the tropics to the poles, both in the atmosphere, through its latent heat, and in the ocean. Water plays a fundamental role in climate.

Water provides key feedbacks to the climate system through evaporation, atmospheric water vapour content, precipitation and the reflectivity of ice, but climate also impacts the hydrological cycle through precipitation and subsequent river discharge. Humans depend strongly on water availability for their food supply. Luckily, water is a renewable resource: the mean residence time of water in the atmosphere is only seven days. The continuous recycling of water through the hydrological cycle means that, globally, we will never run short of water. That is not to say that everyone has access to good-quality water in sufficient amounts all the time. On the contrary, there is often too much in one place and too little in another. In an attempt to reduce this unequal supply, humans have been responsible for considerable changes in the regional water cycle through the damming of rivers and building of reservoirs (Vörösmarty & Sahagian, 2000). The increased world food production is also taking place at the cost of dwindling groundwater resources (Dalin et al., 2017).

The global distribution of water is given in Figure 8.1. By far the largest amount of water is to be found in the world oceans; however, this water is saline, as we have seen in Chapter 7 (Shiklimanov, 2005). The total freshwater store comprises only 2.5% or 34.7×10^6 km^3. Of this, only a mere 1.3% is available in the form of surface water for human use: drinking water, water for food production, industrial use, and so on. Glaciers and ice caps store 68.6% of the world's freshwater in the form of ice, and around 30% resides in our aquifers as groundwater. For climate feedbacks, atmospheric water vapour

Biogeochemical Cycles and Climate. Han Dolman, Oxford University Press (2019). © Han Dolman.
DOI: 10.1093/oso/9780198779308.001.0001

Distribution of Earth's Water

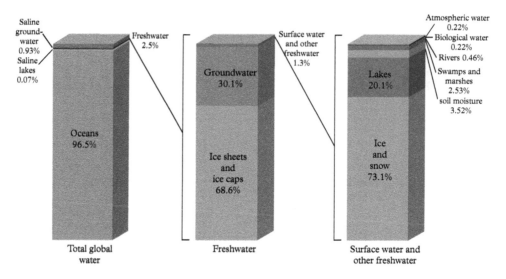

Figure 8.1 *Global distribution of water. Subpartitioning of water stores is from left to right. (After Shiklimanov, 2005)*

Table 8.1 *Average fluxes in the hydrological cycle, obtained by forcing water balance closure and using several satellite products (Rodell et al., 2015); uncertainties are given as the one standard error in the second column after each term.*

	Precipitation (mm/day)		Evaporation (mm/day)		Runoff (mm/day)		Storage (mm/day)	
Land	2.18	0.09	1.32	0.09	0.86	0.08	0.86	0.08
Ocean	3.03	0.17	3.37	0.17	0.34	0.03	−0.34	0.14
World	2.79	0.15	2.79	0.15	0.00	0.02	0.00	0.03

(12,900 km³ in total) is important, as are the ice caps (24.1 × 10⁶ km³) and the ocean (13,338 × 10⁶ km³).

Table 8.1 presents what is arguably one of the most complete current estimates of the global water cycle fluxes (Rodell et al., 2015). These fluxes are based on a large set of satellite data of the previous decades and several data assimilation products.

Importantly, the uncertainties of the various estimates are also given. Table 8.1 shows the fluxes, including those over the ocean, in millimetres per day. (N.b. as is standard practice, we use the hydrologically much used unit mm in this chapter to express the depth of water on a surface, rather than the more correct 10⁻³ m.) Annual precipitation, evapotranspiration and runoff over the global land surface is, in volume units of cubic kilometre according to these authors, estimated to total $116{,}500 \pm 5100$ km^3 yr^{-1}, $70{,}600 \pm 5000$ km^3 yr^{-1} and $45{,}900 \pm 4400$ km^3 yr^{-1}, respectively. Annual precipitation and evaporation over the global ocean surface are then $403{,}500 \pm 22{,}200$ km^3 yr^{-1} and $449{,}400 \pm 22{,}200$ km^3 yr^{-1}, respectively. Advection from the oceanic atmosphere to the land, the surplus of precipitation minus evaporation of the oceans, is a further $45{,}800 \pm 2.1$ km^3 yr^{-1}. The return flow from the land, discharge into the ocean, is the difference between land precipitation and land evaporation and, within uncertainty bounds, is equal to the advected moisture of the oceans, $45{,}900 \pm 4.4$ km^3 yr^{-1}. Note that the uncertainties are generally of the of order 5% and represent significant amounts of water. Hence, different estimates of the global water balance can vary between them, depending on what assumptions are made.

The mean residence time of water in each of the reservoirs depends on the size of the fluxes with respect to the size of the stock (see Chapter 1). Hence, as mentioned the mean residence time of water in the atmosphere is short, only seven days. It is longer in the ocean, ranging from months to years in the surface layer to centuries and thousands of years for deep waters. It is even longer when frozen—the mean residence time in ice caps and sheets depends not only on their size but also on the (frozen) state of the reservoir and accumulation rates. Climate feedbacks from these types of water typically operate at similar timescales, so the water vapour feedback is relevant at short timescales; the ocean climate interaction varies from annual to thousands of years, while the role of ice caps and sheets provides even longer feedback timescales. It is important to note here that impacts of, for instance, melting ice sheets on sea level rise can be felt earlier. It would also appear that the feedback of sea ice on climate may take place faster than previously thought, given the observed exponential decrease of Arctic sea ice in recent years.

8.2 Water vapour, lapse rate and cloud feedback

The distribution of water vapour in the atmosphere is highly variable in space and time, as it is constantly brought into the atmosphere by evaporation, transported horizontally by advection and vertically by convection and depleted by condensation and precipitation. Over the oceans, the absolute humidity is closely linked to the surface air temperature; it changes roughly at the same rate as the saturated vapour pressure from the Clausius–Clapeyron equation (eqn (6.16)). Over land, this situation is more complex, as evaporation and soil moisture show much more variation.

Figure 8.2 shows the global distribution of integrated water vapour as obtained by the European Centre for Medium-Range Weather Forecasting reanalysis system (Dee et al., 2011). The re-analysis uses a single model that assimilates and integrates a variety of in situ and satellite data. The average is from 1979 to 2017. The highest concentrations of water vapour are found near the equator, with the clear northward movement of the ITCZ visible in the band of tropical moisture moving northward from January to July in the Northern Hemisphere. Near Eastern Asia, the effect of the monsoon is visible in July. Note also the difference in Amazonian water vapour content between January and July. Also visible

January

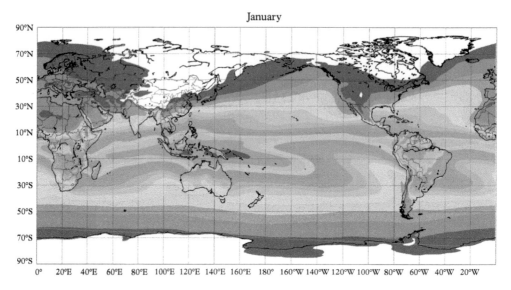

July

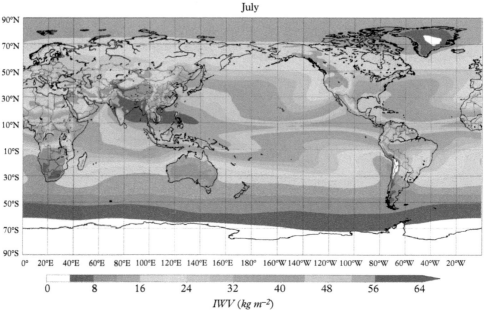

IWV (kg m⁻²)

Figure 8.2 *Mean integrated water vapour content for January (top) and July (bottom) from the ERA-Interim reanalysis of 1979–2017, in kilograms per square metre.*

is the general decrease of water vapour from the equator to the poles. This is a direct consequence of the Clausius–Clapeyron equation (eqn (6.16)) and the declining temperatures in that direction: the atmosphere simply cannot hold more water at lower temperatures. Seasonal variability in atmospheric water vapour is larger in the Northern Hemisphere than in the Southern Hemisphere because of the land-driven temperature differences in the Northern Hemisphere. Water vapour cycles constantly through the atmosphere as it evaporates from the land, condenses to form clouds and eventually returns to Earth as precipitation. Movement of water vapour always incorporates movement of energy (see Chapter 3).

Let us start with the role of water vapour in climate. Most of the atmospheric water vapour resides in the atmospheric boundary layer, the lowest 2 km or so of the atmosphere. If all the water vapour in the atmosphere were to condense, it would amount to a depth of water of only 25 mm. Water vapour is, however, the most critical component of the hydrological cycle, because of its interaction with climate. This is because water vapour influences greenhouse gas absorption: it is the strongest greenhouse gas absorber in the atmosphere, as its absorption bands cover almost the entire infrared region (Chapter 4). Water vapour when ascending and cooling produces cloud. Its abundance has a very strong dependence on temperature through the Clausius–Clapeyron equation (eqn (6.16)). All of this forms the basis of one of the strongest positive feedbacks in the climate system. The water vapour feedback is strongest in the upper troposphere of the tropics due to the combination of high temperatures and low specific humidity. The water vapour feedback thus brings us back to the heart of the greenhouse gas problem.

Let us recall the climate sensitivity equation, eqn (2.3):

$$S^a = \frac{\Delta T}{\Delta R} = \frac{-1}{\lambda_p + \sum_{i=1}^{n}\lambda_i^f} \tag{8.1}$$

In this equation, ΔT is the change in temperature resulting from a change in radiative forcing ΔR, and λ_p is the Planck feedback parameter (the longwave radiation feedback from a pure blackbody; see Chapter 2), the climate feedback to increasing carbon dioxide in the absence of any other feedback The λ_i^f's are the feedbacks associated with other, fast-reacting climate processes. One of these latter is the water vapour feedback; others are lapse rate, cloud and albedo feedbacks. We are dealing here with short timescales only, so feedbacks such as the ice sheet feedback are not included. The water vapour and cloud feedback are intimately associated. But, to be able to determine these feedbacks, one needs climate models that can isolate each of the components separately. Figure 8.3 shows the results of these calculations for several IPCC models that are used for the calculation of feedback factors under high carbon dioxide scenarios (Boucher et al., 2013). The Planck feedback is around −3.2 W m^{-2} °C^{-1}, while the water vapour feedback is around 2 W m^{-2} °C^{-1}. This is thus by far the strongest natural feedback.

The lapse rate feedback is strongly associated with the water vapour feedback. In Chapter 4, we assumed in the explanation of the effect of increasing greenhouse gases in

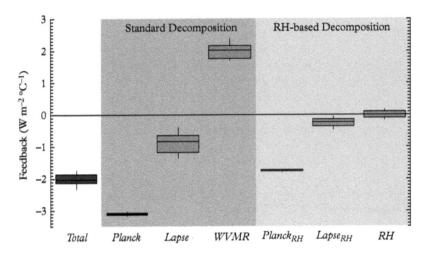

Figure 8.3 *Water vapour feedback parameters associated with lapse rate predicted by CIMP3 global climate models, with boxes showing the interquartile ranges, and whiskers showing extreme values. At the left is shown the total radiative response, including the Planck response. In the darker shaded region is shown the traditional breakdown of this into a Planck response and individual feedbacks from water vapour (labelled 'WVMR') and lapse rate (labelled 'Lapse'). In the lighter-shaded region at right are the equivalent three parameters calculated in an alternative, relative humidity-based framework; RH, relative humidity. (From Boucher et al., 2013)*

the atmosphere that the lapse rate was did not change; by increasing surface temperature due to the greenhouse gas effect, the temperature profile would shift towards higher temperatures, but the slope of the profile would remain the same. If, as we have noted in the tropics, we were to find enhanced warming in the upper troposphere, this would force the lapse rate to change, becoming closer to the moist adiabat (see Chapter 6) that is generally representative of the humid tropical atmosphere. Uneven vertical warming thus leads to a change in lapse rate. Outside the tropics, the warming is expected to be closer to the surface and, there, the lapse rate feedback, λ_1^f, is greater than 0. Globally, the lapse rate feedback is the sum of these various zonal components and, there, λ_1^f is negative ($\lambda_1^f = -0.8$). This is shown in Figure 8.2. In the IPCC models, the water vapour feedback (λ_w^f) and λ_1^f tend to compensate each other: models with a more negative λ_1^f tend to have a more positive λ_w^f.

The third feedback that relates to the water cycle is the cloud feedback. Clouds have the potential to interact with Earth's radiative balance in two ways. They can reflect shortwave radiation due to their white colour and cool the planet, or they can increase longwave (downward) radiation and further warm the planet (see also Chapter 5). The overall cloud feedback is expected to be small with a tendency for a moderately small positive feedback, but this is highly uncertain. Earth system model simulations with 2× carbon dioxide forcing have shown that these model climates respond with a reduction in total cloud amount between about 55° S and 60° N. This reduces the shortwave reflection and implies a positive feedback. They further show an increase in cloud amount poleward of these latitudes, particularly related to decreasing sea ice. The model simulations also, importantly, show an upward shift of cloud tops at nearly every location. Not only does the cloud level rise; the level at which ice is produced or melts also shift towards higher levels. This provides a positive feedback (i.e. warming) as the emission of longwave radiation by the clouds takes place at higher altitude and thus at lower temperatures—Earth traps more of the heat. The density of the clouds, as indicated by their optical depth (Chapter 4), shows an increase poleward of about 40°, and a generally much smaller decrease equatorward of 40° (Zelinka et al., 2012). In summary, the cloud feedback is very complicated, because different cloud properties become important at different altitudes and both positive and negative effects are involved. The shortwave effect is largest in low clouds over the ocean, while the temperature-altitude effect is largest with medium-level and high clouds.

Figure 8.4 shows a summary of the cloud responses from the latest IPCC report. The solid line at the top indicates the tropopause (the upper limit of the troposphere), which is higher in the tropics than at the poles (see also Chapter 6). The processes mentioned earlier, such as rising high clouds, and more polar and less low clouds, are also indicated. Note that a poleward shift of extratropical storms is predicted. This, of course, has potentially a large impact on extreme events in the hydrological cycle. We will come back to this later in this chapter.

Finally, it is important to note that clouds are very heterogeneous in the real atmosphere (something we can just notice by looking up towards a partly clouded sky) and that, in models, large-scale, simplified parametrizations are often used. To be able to resolve clouds at the right scale in a climate model, a drastic increase in resolution, both in the horizontal and in the vertical, would be needed.

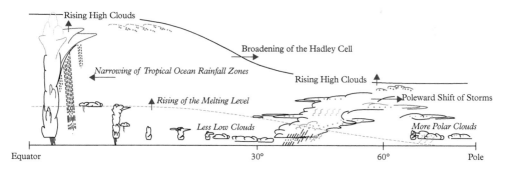

Figure 8.4 *Cloud responses to greenhouse gas warming. Changes anticipated in a warmer climate are shown by arrows, with changes indicated in regular letters indicating those making a robust positive feedback contribution, and those indicated in italic letters indicating those where the feedback contribution is small and/or highly uncertain. No robust mechanisms contribute negative feedback. Changes include rising high cloud tops and melting levels; increased polar cloud cover and/or optical thickness (high confidence); broadening of the Hadley cell and/or poleward migration of storm tracks; narrowing of rainfall zones such as the Intertropical Convergence Zone (medium confidence); and reduced low-cloud amount and/or optical thickness (low confidence). Confidence levels from IPCC. (From Boucher et al., 2013)*

8.3   Transport of atmospheric moisture

The zonal and meridional flows of moisture determine the availability of water vapour for precipitation. Most of the zonal moisture transport over the globe takes place via the easterlies and westerlies we outlined in Chapter 6 (see Figure 8.5; Dee et al., 2011).

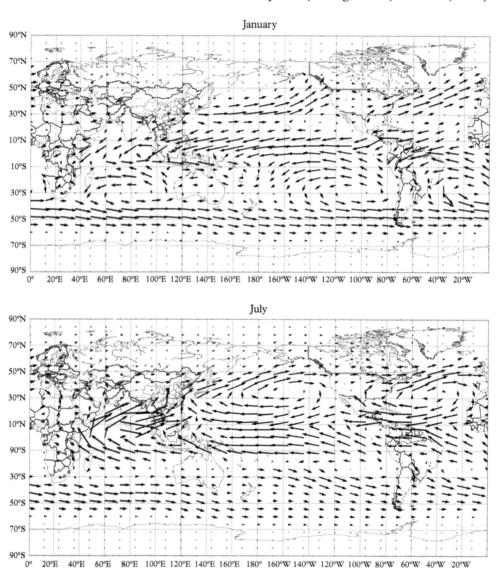

Figure 8.5 *Column-integrated moisture flow from the ERA-Interim reanalysis of 1979–2017, in kilograms per metre per second; the inset vector is 100. The top panel is for January; the lower panel is for July.*

Note that these moisture fluxes are almost two orders of magnitude larger than the meridional fluxes. The result is that the moisture transport at temperate latitudes generally takes place in a west-to-east direction, corresponding to the main prevailing winds. In contrast, in the tropics, the trade winds blow from the east and transport the moisture to the west. This overall picture is modified by the presence of the monsoons, which are seasonally dependent. In July the summer Asian monsoon produces a strong northward flow over the South East Asian continent.

The averaged meridional transport is considerably less. A substantial fraction of the total meridional moisture transport is performed by so-called atmospheric rivers, narrow corridors of high water vapour content that transport water vapour at high speed.

Indeed, for the meridional transport in middle latitudes, atmospheric rivers are estimated to account for 90% of the transport, while they take up only 10% of the space (Gimeno, 2014). The filamentary structure of the rivers from which they derive their name is clearly seen when, for instance, the integrated water vapour content of the atmosphere is plotted (see Figure 8.6). They occur often with extratropical storms and are associated with the atmospheric bombs we discussed towards the end of Chapter 6. The latent heat released by the condensation of water vapour in the river provides the source of energy that is required to sustain the rapid decline in pressure that is typical of the bomb. The magnitude of the moisture flux in a typical atmospheric river is about 1.6×10^8 kg s^{-1}, which is similar to the flux in the Amazon River. The zonal average of the annual mean meridional water vapour flux moving poleward across 30° N is about

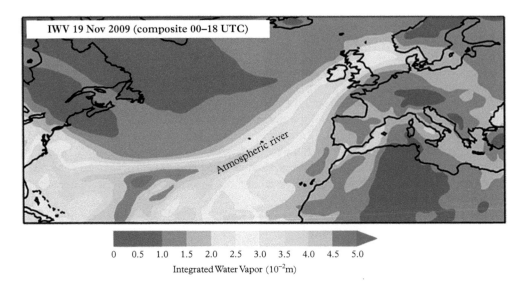

Figure 8.6 *Integrated total volumn of integrated water vapour (IWV) content between 00 and 18 UTC on 19 November 2009, showing the existence of an atmospheric river that produced extreme rainfall in Cumbria, United Kingdom. (Data: ERA-Interim, after Gimeno, 2014)*

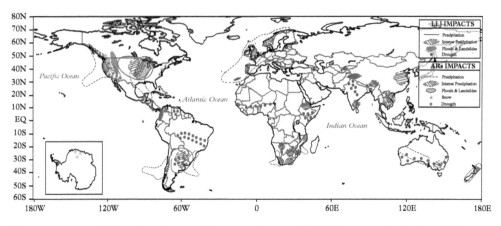

Figure 8.7 *Compilation of the impacts of extremes that can be linked to atmospheric rivers (AR impacts; blue) and low-level jets (LLJ impacts; red) in terms of precipitation (dashed outline), intense precipitation (diagonal stripes), floods and landslides (solid colour), snow (asterisks) and drought (dots). (Figure courtesy of Dr. L. Gimeno, University of Vigo, Ourense, Spain)*

7×10^8 kg s^{-1}. Hence, only four or five atmospheric rivers in a hemisphere can carry the majority of the global meridional vapour flux.

Atmospheric rivers are often associated with extreme rainfall and the resulting flooding, such as occurred in Cumbria in the United Kingdom on 19 November 2009. The atmospheric river seen in Figure 8.6 kept on supplying moisture to precipitation, resulting in a record rainfall—in excess of 300 mm in 24 hours. This inevitably led to severe flooding, with estimated damages of the order of several hundred millions of euros. Figure 8.7 (Gimeno et al., 2016) shows an overview of the areas on the globe where extremes are affected by such atmospheric rivers, which occur mostly in the extratropics, and their tropical counterparts, the low-level jets. The areas that are affected by atmospheric rivers in western Europe and the north-west coast and the mid-eastern region of the United States clearly stand out. In the tropics, we note the south-east part of Brazil and the areas affected by monsoons. Some of these areas can be affected by both extreme drought and extreme rainfall, while others, such as some areas in Australia, are generally more prone to drought impacted by the variability of the atmospheric moisture transport (Gimeno et al., 2016).

8.4 Precipitation

The global mean annual precipitation is 2.67 mm day^{-1}. In the tropics, this is close to 6 mm day^{-1}, while at mid-latitudes it is closer to the global mean, but it decreases towards values close to 0.4 mm day^{-1} at the poles. There is an obvious similarity between the pattern of water vapour in Figure 8.2 and that of precipitation in Figure 8.8. The tropical

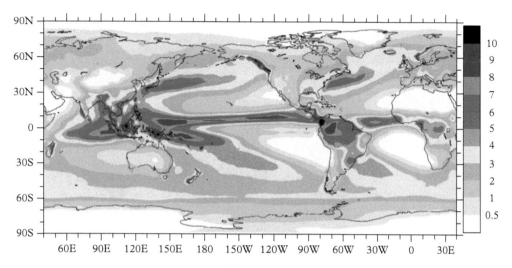

Figure 8.8 *Climatological annual mean precipitation (millimetres per day) for 1979–2010. The areal mean for the entire globe is 2.67 mm day⁻¹. (Data from Climate Data Guide: https://climatedataguide.ucar.edu/climate-data/gpcp-monthly-global-precipitation-climatology-project, accessed 27-3-2017)*

rainfall bands around the equator, and the high precipitation around the maritime continent, are clearly visible. Hotspots of rainfall are to be found near coasts with high orography, such as Lloró, Colombia, and Mount Cameroon, Cameroon. Annual rainfall at these places can be in excess of 10,000 mm. The islands of the maritime continent (e.g. Indonesia, Malaysia) receive as much rainfall as the far larger continent of Australia. Specific conditions on these islands generate large convective activity. It is likely that sea breezes in a moist environment originate at each coast and meet towards the middle of these islands to produce strong convergence and convection. The size of these islands is favourable to such a phenomenon, while, of course, in general, the region remains one of high convective activity per se.

Average long-term precipitation over the land is estimated at 790 mm yr⁻¹. Over the ocean, this is closer to 1100 mm yr⁻¹, but both estimates have considerable year-to-year variability and, over the ocean, the estimates are subject to uncertainties in the satellite retrieval. Over land, rainfall from the synoptic rainfall reporting network can be used for correction and validation.

8.5   Runoff and river discharge

Over land, 39% of the precipitation becomes runoff, but this is very unevenly distributed (Milliman & Farnsworth, 2009). The uneven distributions of precipitation and topography

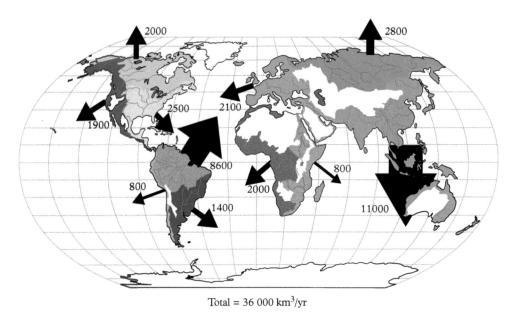

Total = 36 000 km³/yr

Figure 8.9 *Discharge of fresh water to the global coastal ocean. Numbers are the mean annual discharge (in cubic kilometres per year). (After Milliman & Farnsworth, 2009)*

that influence river flows globally lead to a non-uniform pattern in global runoff. Figure 8.9 shows the global distribution of annual discharge into the ocean. Note that the total discharge sum in this figure, 36,000 km³ yr⁻¹, is smaller than the previously quoted 45,900 km³ yr⁻¹. This is a rather large difference, but it is best taken as an indication of our low skill at determining the global quantity of river discharge. For a considerable number of rivers (about 40%), no data are available and these missing data have to be estimated. The specifics of this gap filling determine to large extent the variability between the global estimates. This situation also means that it is still very hard to determine whether the global river discharge has changed as a result of climate change. Locally, if there are good, reliable measurements available, it may be possible to show such a change. Arctic rivers may show an increase in runoff, while rivers in the more semi-arid regions generally appear to show a decrease.

The largest runoff is that by the Amazon River into the North Atlantic. The outflow from the Arctic is relatively low, compared to the drainage area (20% of the global land surface). Here, the low annual precipitation is the main cause of the low runoff. The runoff into the Arctic Ocean has, however, the potential to make a strong impact on climate—changes in runoff would alter the freshwater balance by decreasing salinity and ultimately affect the formation of deep water (see Chapter 7). High-runoff catchments are located predominantly in northern South America, South East Asia and the maritime continents of Oceania. Twelve rivers individually discharge more than 400 km³ yr⁻¹. Although these rivers occupy only 25% of the total land area emptying into the global ocean, together they account for more than 35% of the total water that is annually discharged into the ocean.

There is considerable temporal variation in river discharge, at scales from seasons, to years, to decades and even longer. Seasonal variability often has to do with the occurrence of rainfall seasons that produce marked changes in runoff. At annual scales, the variability appears driven by several climate teleconnections, such as El Niño, the Arctic Oscillation, the Pacific Decadal Oscillation, the Atlantic Meridional Oscillation and the Southern Annual Mode. Given the strong dependence of rainfall on El Niño in the tropics, most of the tropical rivers show discharge patterns that reflect the variability in El Niño Index.

These climate patterns are often described by simple indices, such as the differences in the anomalies of sea level pressure between two locations. For example, the difference between pressure on Tahiti (East Pacific) and at Darwin (West Pacific) defines the Southern Oscillation Index, which is very much tied to the El Niño–La Niña sequence. El Niño events involve large exchanges of heat between the ocean and atmosphere and affect global mean temperatures. At the time, the 1997–8 event was the largest on record in terms of sea-surface temperature anomalies, and the global mean temperature in 1998 was the highest on record, based on a climatological mean of thirty years. It was thought to have contributed 0.17 °C to the warming. The 2015–16 El Niño is thought to have driven the global temperature to its then maximum value. The contribution of El Niño to that maximum was estimated to be 0.1–0.2 °C. From a hydrological perspective, the associated changes in rainfall are equally important. El Niño increases flooding in some areas and is held responsible for massive droughts in others.

El Niño arises when the trade winds from the east start to weaken (see Chapter 6). This sets in motion a complex chain of events, involving warming of ocean waters near the coast of Peru and associated changes in ocean circulation. The most visible expression is a change in sea-surface temperature in the Pacific (see Figure 8.10). The movement to the east of the hot pool brings with it a move of the main area of convective activity, leaving the maritime continent drier than usual. Figure 8.10 shows the sea-surface temperature in the build-up of the 2015–16 El Niño event in March 2015, and the very warm pool of ocean water off the coast of Peru in December 2015. No wonder El Niño, which translates as 'The Child', is named from its occurrence near Christmas.

River discharge can have a significant impact on the hydrography and biogeochemistry of the oceans. Nutrients are delivered through natural leaching and erosion processes in the upstream catchments and delivered by runoff, the precipitation minus evaporation excess, to the coastal ocean. There, they impact the biogeochemistry of the coastal ocean at short timescales. At geological timescales, river discharge and transport of materials have an influence on the seawater composition and salinity (see Chapter 7).

March 2015

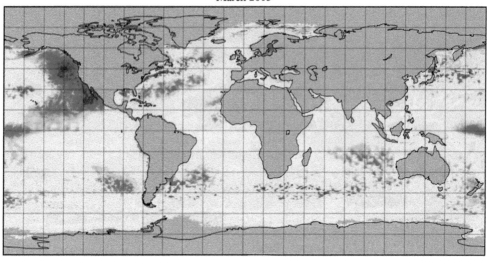

December 2015

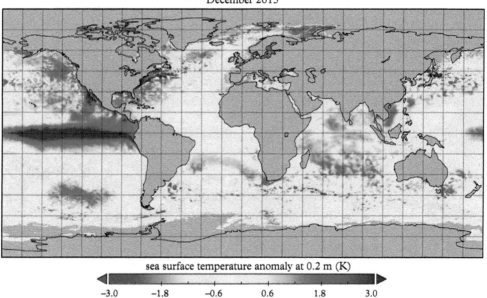

sea surface temperature anomaly at 0.2 m (K)

-3.0   -1.8   -0.6   0.6   1.8   3.0

Figure 8.10 *Sea-surface temperature (SST) records from a combination of satellite data, showing the build-up of a warm pool near the coast of Peru in 2015. The upper panel is for March; the lower panel is for December. (ESA-CCI, accessed 3-4-2017: http://gwsaccess.ceda.ac.uk/public2/nceo_uor/sst/L3S/ EXP1.2/plot/monthly/)*

8.6   Evaporation

Evaporation from the world's oceans constitutes the largest component of the global water balance, with a total flux of roughly 450,000 km^3 yr^{-1}. It is important as a source of moisture that is tied not only to the radiation absorption processes of water vapour but also to the availability of freshwater, which governs the habitability of the planet. The atmosphere advects the excess of evaporation over precipitation over the oceans to the land, where it contributes to the precipitation excess over evaporation that ultimately returns to the ocean as river runoff. Over the ocean, the balance between precipitation and evaporation is a key factor in determining salinity. Over land, the balance between precipitation and evaporation determines the availability of water and the occurrence of drought.

Climatologies of land and ocean evaporation can be obtained by using satellite data and reanalysis products (Bentamy et al., 2017; Mueller et al., 2013; Yu & Weller, 2007), but no direct observation at sufficiently large scale exists. Figure 8.11 shows the results of such a reanalyses product for the ocean flux, based on the years 1979–2017 (Dee et al., 2011). The annual mean pattern of ocean latent heat flux (latent heat flux = λE, where λ the latent heat of vaporization, and E is evaporation) reflects the dominant climate and ocean current features of the two hemispheres and suggests a largely bimodal pattern in the latent heat loss of the oceans. The largest latent heat fluxes in January are associated with the two western boundary currents (Chapter 7): the Gulf Stream, and the Kuroshio along the Japanese archipelago and its extensions. These surface currents play an important role in transporting tropical heat poleward and, in boreal winter, they release large amounts of latent and sensible heat to the atmosphere. In July, the largest fluxes are located over the broad subtropical southern Indian Ocean and the boundary regions associated with the Eastern Australian Currents. Ocean latent heat flux is relatively weak over the eastern Pacific and Atlantic cold regions, and also at high northern and southern latitudes.

Figure 8.12, in contrast, shows evaporation of the land from a model that uses microwave remote sensing observations of soil moisture to constrain potential evaporation rates (Martens et al., 2017). The four panels show evaporation from forest rainfall interception loss, evaporation from soil, transpiration and total evaporation. For forests, it is important to make the distinction between interception evaporation and transpiration. Interception is the process whereby a tree canopy intercepts rain which then evaporates directly from the wet surface. Particularly in forests, this is an important process that can contribute up to 20%–40% of the evaporation locally. Transpiration is the plant physiological process by which water is taken up from the soil into the root system and transported via the stems and leaves to finally exit the plant through the stomata as water vapour. This process depends on the behaviour of the stomata, which are known to respond to factors in their environment, in particular, radiation, temperature, vapour pressure deficit and soil moisture. Stomata also play a key role in the carbon cycle, as they take up carbon dioxide when they release water (see Chapter 9). Transpiration is largest in wet, humid regions that have no season. In temperate regions, in summer,

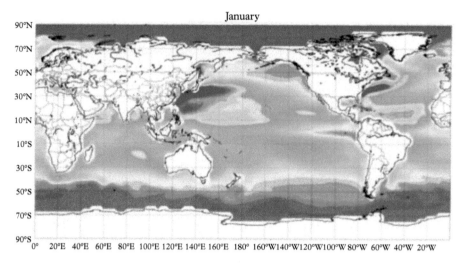

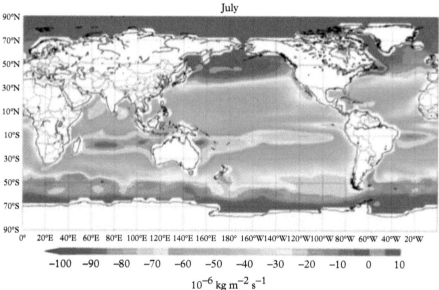

Figure 8.11 *Mean moisture flux (evaporation) over the ocean from an ERA-Interim reanalysis of 1979–2017, in kilograms per metre per second. The upper panel is for January; the lower panel is for July. (Dee et al., 2011)*

deciduous trees contribute substantially; in winter, the only evaporation comes from soil evaporation and interception. According to this well-validated model, the total land evaporation amounts to 66×10^6 km^3 yr^{-1}, with 74% due to transpiration, 15% due to bare soil evaporation (including snow sublimation) and 11% due to interception evaporation. These numbers depend slightly on the precise formulation of the model and

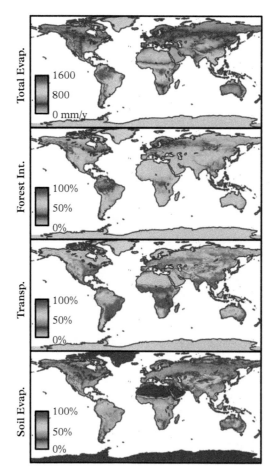

Figure 8.12 *Evaporation over land for 2011–15, showing soil evaporation, transpiration, forest interception loss and total evaporation. (From Martens et al., 2017)*

which input data are used but, from other observations as well (using water isotopes), it would appear realistic to say that transpiration contributes to about 75% of the total evaporation from land.

Evaporation over land is controlled by the amount of moisture in the soil. If there is no moisture, as for instance in deserts, there will be no evaporation. However, when there is moisture, variability in precipitation, for instance induced by one of the climate indices we saw previously, the Southern Oscillation, may also cause the soil moisture to vary and hence so will the evaporation. Over oceans, decadal trends in the sea-surface temperature will also provide variability in evaporation.

Figure 8.13 shows the general upward trend in evaporation as calculated by the same as used in Figure 8.12, that uses satellite microwave data for soil moisture, surface temperature and vegetation optical density (Miralles et al., 2014). This trend is expected

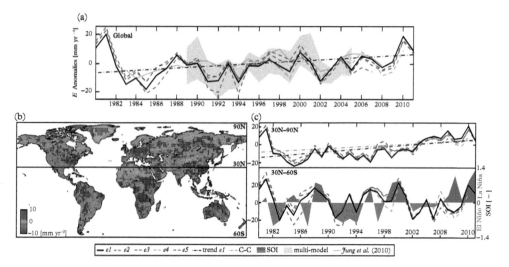

Figure 8.13 *Interannual variability in (a) simulated anomalies of global land evaporation, (b) anomalies in land evaporation and (c) anomalies in southern and northern extratropical lands. Also shown in Panel (a) are several small modifications in the algorithms (e1–e5) and the expectation from the Clausius–Clapeyron equation. The lower component of Panel (c) also shows the variability in the Southern Oscillation Index (SOI) in the Southern Hemisphere. (After Miralles et al., 2014)*

because evaporation is expected to increase with higher temperatures. There are also indications that the trend is not robust, as several periods appear to show stagnation. When this global trend is split to give separate Northern and Southern Hemispheric variability, we notice that the variability in the Northern Hemisphere roughly follows what would be expected from the Clausius–Clapeyron equation (eqn (6.16)), while the variability in the Southern Hemisphere is more pronounced and co-varies strongly with the Southern Oscillation Index. This is caused by the fact that, in the Southern Hemisphere, during El Niño years there is less rainfall and consequently less moisture to evaporate. In contrast, during La Niña years, the reverse happens and evaporation is above average. The combination of the southern and northern responses provides a global response that is lower than expected and more variable. It is worth noting, though, that, even with surface temperatures rising, there will be periods of drought and reduced evaporation.

8.7   Recycling of moisture

When moisture evaporates, it enters the atmospheric pool, where it can again contribute to precipitation. Several studies have estimated this amount of recycling on different continents and show the relevance of recycling, particularly for continental areas far from the ocean. For this purpose, we divide precipitation over a continent into two components: the precipitation that is derived from evaporation over land, and that coming from the ocean. This reasoning leads to the following two recycling ratios (van der Ent et al., 2010):

$$\Omega_c = \frac{P_c}{P} \tag{8.2a}$$

$$\Psi_c = \frac{E_c}{E} \tag{8.2b}$$

where eqn (8.2a) gives the continental precipitation recycling ratio, and eqn (8.2b) gives the evaporation recycling ratio. The continental precipitation recycling ratio shows the dependence of precipitation at a certain location and time on the contribution from upwind continental evaporation P_c to sustain the total precipitation P on land. The evaporation ratio shows how much evaporation at a certain location contributes to downwind precipitation. The data obtained from these two ratios are shown in Figure 8.14. China and central Asia, the western part of Africa, including the Sahel zone, and central South America show high recycling ratios, indicating that most of the precipitation there

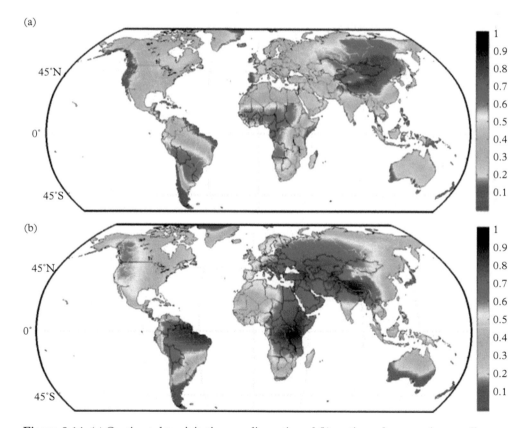

Figure 8.14 *(a) Continental precipitation recycling ratio and (b) continental evaporation recycling ratio from reanalysis data from 1999 to 2008. (After van der Ent et al., 2010)*

is of terrestrial origin (Figure 8.14(a)). The evaporation recycling ratio is high on the North American continent, in the entire Amazon region, in central and east Africa and in a large area in the centre of Eurasia. These are the continental moisture sources of the world. In contrast, the areas that are major sinks for continental evaporated water (Figure 8.14(b)) are the north-east of North America, the region around the savannah belt of South America, central and west Africa and large areas of China, Mongolia and Siberia.

On average, about 40% of the terrestrial precipitation originates from land evaporation, and about 57% of all terrestrial evaporation returns as precipitation over land. The evaporation–precipitation feedback is central to the hydrological cycle. Importantly, this feedback may change, as both evaporation and precipitation show sensitivity to climate change. In particular, it plays a role in large continents where forests are being converted to agricultural use, such as in Brazil. Lower albedo, deeper roots and aerodynamic roughness of rainforest in comparison to grass or agricultural crops generally lead to higher evaporation from forest. This evaporation enters the atmosphere, where it contributes to rainfall. Estimates of the recycling in Amazonia are that up to 30%–40% of the precipitation comes from previously evaporated moisture. Now, if, through deforestation, this amount of evaporation decreases, rainfall potentially decreases as well; the size of the reduction will depend on the recycling ratio and evaporation decrease. The topography of South America furtermore creates a particular situation: the Andes Mountains to the west of Amazonia block the westward moisture flow, which turns south after hitting the Andes. It is thought that reduced evaporation in the deforestation zones in the east may thus affect rainfall further south. Research into the specifics of these transport mechanisms of moisture and the effect of deforestation is relatively recent.

8.8   Frozen water

Of the global fresh water resources, 68.6% consists of frozen water in the form of ice caps and glaciers. Collectively, areas with frozen water or snow make up the 'cryosphere', a term originating from the Greek word *kryos*, which means 'cold'. The different components of the cryosphere are depicted in Figure 8.15.

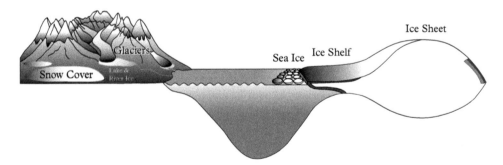

Figure 8.15 *Components of the cryosphere. (After Vaughan et al., 2013)*

The Antarctic and Greenland ice sheets, at present, cover 8.3% and 1.2% of the global land surface, respectively; glaciers cover another 0.5%. Seasonally frozen land and permafrost (continuously frozen land) cover a further 33% and 9%–12%, respectively. This indicates that almost 50% of the land surface area is, in one way or another, part of the cryosphere. Of these, of course, only the ice sheets, ice caps and glaciers would have a significant impact on global sea levels if they were to disappear through melting. Expressed as sea level equivalents, these components can potentially contribute 58.8 m, 8.36 m and 0.41 m, respectively. Most of the ice on the ocean is sea ice, and this covers a maximum of 5.2% and 3.9% of the ocean surface for the Antarctic and Arctic sea ice, respectively (Vaughan et al., 2013).

The most important frozen-water feedback on the climate is the ice–albedo feedback; in fact, this may be the most important climate feedback of all. The albedo feedback not only concerns sea ice, but also ice caps and sheets, through their albedo. In Chapter 1, we referred to periods when most or all of Earth was covered by ice: Snowball Earth. The first period, the Huronian glaciation, was 2.4–2.3 Gyr ago. In that case, declining methane levels are thought to have been the possible culprit. Later, for the Sturtian glaciation, 710 Myr ago, enhanced carbon dioxide uptake through weathering (eqns (1.2a, b)) is thought to have played a major role in bringing down temperatures, much as in the last Snowball Earth, the Marinoan glaciation, 635 Myr ago. There is considerable debate as to whether, during these periods, all of Earth was covered with ice and to what depth the oceans were frozen. These Snowball Earth periods were short compared to the hothouse conditions that ruled for most of the geological past. During the Pleistocene, we have experienced glacial–interglacial cycles where glacials, periods of enhanced ice sheet growth, were interchanged with warming periods, the interglacials. We will not go further into these glaciations, exciting as they are, as a whole literature exists on all the aspects, including the potential role of greenhouse gases when exiting a glacial (e.g. Alley, 2000).

Sea ice extent shows a clear seasonal cycle. It forms in the winter, and parts of it thaw during summer. Sea ice observations from satellites show that thawing has increased over the past ten to twenty years in the Northern Hemisphere, leading to relatively low extents in September. Next to extent, the depth of sea ice is also important. While the extent can be observed with satellites, measuring ice depth is problematic. Nevertheless, thermodynamically, the volume matters, as this is what determines the latent heat content of the ice.

Figure 8.16 shows what is known as the 'Arctic Death Spiral'. It shows how both the summer sea ice volume and the winter sea ice volume have declined over the satellite observing period. The black line from the model is the September volume. This volume was 17×10^3 km^3 in 1979, when measurements started, but is now less than 5×10^3 km^3. Before 1979, observations are scarce, but a recent compilation of various regional sources, including logbooks from whaling ships, has made an extension back to 1850 possible. This compilation showed that the current rate of decline is unprecedented.

The ice–albedo feedback from the sea ice works so strongly because the difference in albedo between the sea ice and the open water is large. When warmer weather or higher ocean temperatures cause more sea ice to thaw, the extent of fresh water increases and, due to its low albedo, this further warms up the lower atmosphere. This positive feedback

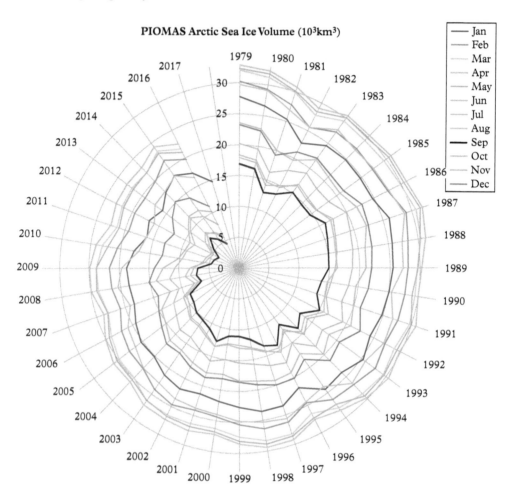

Figure 8.16 *Sea-ice volume calculated from extent and model data. Monthly averages from January 1979 to January 2017. The several lines refer to months of the year. Data: http://psc.apl.washington. edu/wordpress/research/projects/artic-sea-ice-volume-anomaly/. © 2017 Andy Lee Robinson (After PIOMAS, http://psc.apl.uw.edu/research/projects/arctic-sea-ice-volume-anomaly)*

is thought to be responsible for the accelerated decline in Arctic sea ice. It is thought that the change in sea ice extent has implications for the behaviour of the jet stream, and even changes in thermal wind patterns have been linked to changes in Arctic sea ice. Generally, it is thought that a wavier jet stream will result from changes in sea ice, creating the potential for increased impacts on mid-latitude extreme weather. Antarctic sea ice behaves differently and, up until recently, was extending in area.

Ice caps and ice sheets contain by far most of Earth's fresh water (Figure 8.1). Their impact on society lies particularly in their linkage with the ocean through the supply of water and the consequent contribution to sea level rise. However, in past climates, particularly since 800,000–1,000,000 years ago, they have been major players in controlling climate. We saw in Chapter 2 how Earth has 'landed' in a glacial–interglacial cycle with a roughly 100,000-year recurrence time (Figure 2.3). The impact on sea level of the large ice sheets is enormous (Clark et al., 2009). The right panel in Figure 8.17 shows the extent of the ice sheets during the period of the Last Glacial Maximum, roughly 25,000 years ago. The extent of the large ice caps over North America—the Laurentide Ice Sheet and the Cordilleran Ice Sheet—are clearly visible, as well as the ice caps over Northern Europe and the British Isles (marked SIS and BISS, respectively). To appreciate how much water is involved, the corresponding change in relative sea level (marked RSL on the figure) is plotted in Part (C) in the left panel. According to this graph, at the Last Glacial Maximum, the global sea level stood 145 m lower than it does today. For comparison, if today's Antarctic ice sheets were to melt, a sea level rise

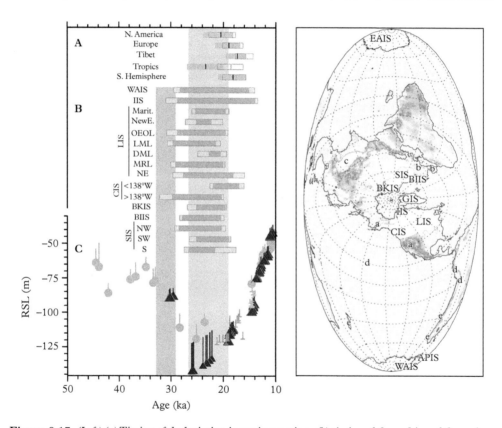

Figure 8.17 *(Left) (a) Timing of deglaciation in various regions, (b) timing of the melting of the main ice caps and sheets and (c) the corresponding relative sea level (RSL) change. (Right) Distribution of ice sheets during the Last Glacial Maximum; APIS, Antarctic Peninsula Ice Sheet; BIIS, British Isles Ice Sheet; BKIS, Barents–Kara Ice Sheet; CIS, Cordilleran Ice Sheet; EAIS, East Arctic Ice Sheet; GIS, Greenland Ice Sheet; LIS, Laurentide Ice Sheet; SIS, Scandinavian Ice Sheet; WAIS, West Antarctic Ice Sheet. The letters a–e denote areas with mountain glaciation. (After Clark et al., 2009)*

of 58.3 m is foreseen, while the corresponding estimate for the Greenland ice sheet is 8.36 m. Melting of the glaciers would contribute a further 0.42 m. To give an idea of the volume of water involved, it requires 361.8 km^3 of water to raise the global sea level by 1 mm—so we are talking considerable quantities of water.

The effects of these considerable quantities of ice are visible in the isotope records of oxygen and make an excellent tracer of ice volume. As we have seen before, the calcareous skeletons of foraminifera are built with oxygen taken from sea water. The isotopic composition of the skeleton reflects that of the sea water at the time of uptake and the temperature. Normally, evaporation produces a known change in the ratio of the heavier ^{18}O isotope and its lighter variant, ^{16}O with more ^{16}O in the vapour phase leaving the liquid water enriched in ^{18}O.

Ice, however, is generally depleted of ^{18}O, compared to the ocean. Here, processes happening in the clouds during condensation are important: during condensation, the vapour gets depleted of ^{18}O (and ^2H), as the heavy isotopologues rain out. During warm periods, the air reaching the poles still has heavy water isotopes when it forms precipitation (snow) over the poles. In contrast, during glacials, the air masses reaching the pole are much colder and, since much of the vapour has already condensed and precipitated out, the remaining vapour is depleted of ^{18}O. Therefore, ice deposited during cold periods is depleted in ^{18}O (and ^2H), and that deposited during warm periods is enriched in these isotopes, unlike what is seen in the sediment record.

If the ice masses with the lighter isotopes were to melt, they would release huge quantities of $H_2{}^{16}O$ into the ocean. This changes the isotope ratio of oxygen in water and thus provides a convenient proxy for ice storage. There are differences in the isotope composition of planktonic versus benthic foramnifera shells, with the latter storing more ^{18}O as a result of their deep ocean environment. It has been suggested that ice volume varies with isotope ratio at 1‰ per 10 m sea level. In terms of volume of ice, for the Last Glacial Maximum, this would be equivalent to in excess of 55×10^6 km^3 ice, compared to today's volume of 36×10^6 km^3.

9

The Carbon Cycle

9.1 Carbon dioxide variability at geological timescales

The large-scale time dynamics of the carbon cycle can best be illustrated by the geological scale variability in atmospheric carbon dioxide. Figure 9.1 shows a reconstruction of carbon dioxide based on several proxies, such as stomata, liverworts and phytoplankton, which all show responses to atmospheric carbon dioxide concentration and thus allow, to some uncertainty, reconstruction of the past concentration (Beerling & Royer, 2011). The figure shows that atmospheric carbon dioxide levels have varied enormously, from values as high as 6000 ppm to values around 200 ppm in more recent geological times. What makes this figure even more interesting, next to the overall long-term drop in atmospheric carbon dioxide, is the correspondence of lower values with geologically well-documented ice ages. In Chapter 2, we saw how this has held for the last 900,000 years, but the correspondence between low carbon dioxide and cool temperatures or ice ages also holds up for the past 600 million years. Very low values of carbon dioxide are encountered at 300 Myr ago, during the Permian–Carboniferous ice age, and the late Cenozoic (30 Myr ago), with high values (>1000 ppm) at all other times. Only in the late Ordovician (~440 Myr ago) do we find a carbon-dioxide-rich atmosphere with glacial conditions coexisting. Towards the last 0.9 Myr ago, the values vary between 280 and 180 ppm (Chapter 2). So, the general picture from the paleo data is that low carbon dioxide levels are associated with cool conditions or ice ages, and high carbon dioxide levels are associated with warm conditions.

Figure 9.1 illustrates the increases in atmospheric carbon dioxide concentrations that we discussed in Chapter 2, primarily those at geological timescales. These increases reflect emissions from tectonic processes, such as at subduction zones, though volcanoes and mid-oceanic ridges (see Chapter 1). Changing those, for instance in a model representation of carbon dioxide evolution, has a significant impact on the carbon dioxide concentrations that are simulated in a model (Van Der Meer et al., 2014). In this chapter, we will further investigate how the various components of Earth's carbon budget work together to produce these changes, and how we can relate this to climate at various timescales.

Before going to these issues, we need to appreciate the current distribution of carbon in Earth's main surface reservoirs (Figure 9.2). Remember how we showed in Chapter 1

Biogeochemical Cycles and Climate. Han Dolman, Oxford University Press (2019). © Han Dolman.
DOI: 10.1093/oso/9780198779308.001.0001

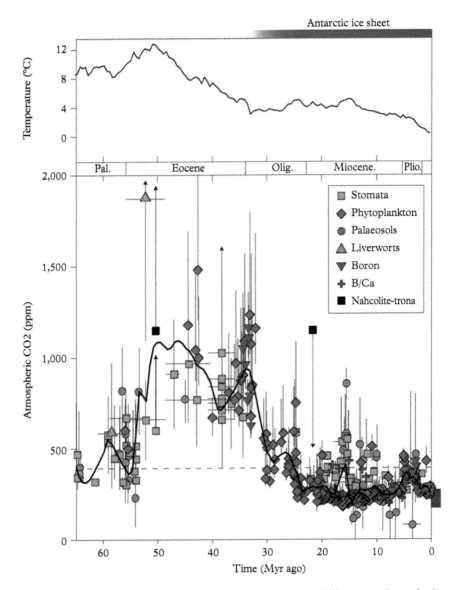

Figure 9.1 *Evolution of atmospheric carbon dioxide over the last 65 million years. Atmospheric carbon dioxide reconstructed from seven proxies. The upper panel shows the temperature evolution; Olig., Oligocene; Pal, Paleocene; Plio, Pliocene. (After Beerling & Royer, 2011)*

that these comprise only 1%–2% of the carbon stored in the deep Earth and crust. By far, the largest amount of carbon in the surface reservoirs is locked up in the carbon stock of the intermediate and deep ocean: 37,100 Pg C. Compared to that very large stock, the other amounts are much smaller. In the atmosphere, where the key radiative effects are cause of concern, only a minor fraction is found—approximately 600 Pg C. Vegetation

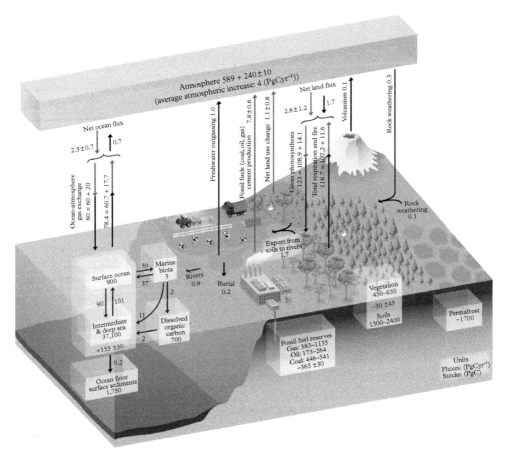

Figure 9.2 *Stocks of carbon, in petagrams of carbon, in the pre-industrial world (before 1850; black text), with the impact from more recent activities such as fossil fuel burning and deforestation indicated in grey text. The soil component includes 1700 Pg C locked up in soil in permafrost areas. Note that part of the fossil fuel that is burnt enters the other reservoirs. (After Ciais et al., 2013)*

and soil lock up a further amount: 1950–3050 Pg C. Fossil fuel reserves are estimated to contain 1002–1940 Pg C. Exchanges between the various reservoirs and adequate estimation of stocks and anthropogenic fluxes are key to understanding the impact of adding the fossil fuel carbon to the other reservoirs.

This chapter first describes the key aspects of the ocean and terrestrial carbon cycle, subsequently indicating their roles in the variability of the geological carbon cycle, in the glacial–interglacial cycles and, finally, in today's human perturbation of the carbon cycle. It ends by discussing possible future scenarios for the carbon cycle's response to the current human perturbation.

9.2 The ocean carbonate system

The ocean carbon cycle provides the bridge between the long-term, geological, tectonically driven carbon cycle and the short-term, biologically driven carbon cycle. As we have seen, ocean sedimentation of carbonate sediments is eventually reorganized through the tectonic cycle to provide new carbon dioxide to the atmosphere (in common with a variety of other gases). The key chemical reactions of the ocean are given by the dissolution of carbon dioxide in water and the formation of bicarbonate (hydrogen carbonate):

$$CO_2 + H_2O \rightleftarrows HCO_3^- + H^+ \tag{9.1a}$$

$$HCO_3^- \rightleftarrows CO_3^{2-} + H^+ \tag{9.1b}$$

The equilibrium of these chemical reactions depends critically on the pH of the seawater. Low pH, that is, a more acidic ocean, will tend to push both the reactions to the left, with less uptake of carbon dioxide, whereas high pH will push them towards the right and the production of carbonate. At typical seawater values of pH = 8.2, this would lead to an equilibrium distribution of 0.5% dissolved carbon dioxide, 89.0% carbonate and 10.5% bicarbonate (Zeebe, 2012). Figure 9.3 shows the behaviour of these fractions as a function of the pH, indicating the strong sensitivity of the ocean's carbonate system to ocean acidification. The sum of these values in gravimetric units (e.g. micromoles per kilogram) is called total dissolved inorganic carbon (TCO_2), also known as dissolved inorganic carbon.

Another important quantity in ocean biogeochemistry is total alkalinity. This expresses the excess of proton acceptors (bases):

$$TA = [HCO_3^{2-}] + 2[CO_3^{2-}] + [OH^-] + [B(OH)_4^-] - [H^+] \tag{9.2}$$

where *TA* is total alkalinity. With these two quantities—TCO_2 and total alkalinity—it is possible to fully describe the ocean carbonate system. This system is sensitive to temperature and salinity changes. Hence, colder waters take up carbon dioxide and have higher TCO_2, and warm tropical waters tend to outgas carbon dioxide, with a consequently lower TCO_2. Warm regions thus tend to have higher carbonate concentrations and are more saturated with respect to carbonate than colder regions. The carbonate system is, in this way, a precise indicator of (past) climate conditions.

The ratio of the change in carbon dioxide to dissolved inorganic carbon is called the Revelle factor and can be expressed as

$$R = \frac{\partial \ln[CO_2]}{\partial \ln DIC} \tag{9.3}$$

where *DIC* stands for dissolved inorganic carbon (Egleston et al., 2010). This factor is named after Roger Revelle, who was, together with Hans Suess, the first to realize that

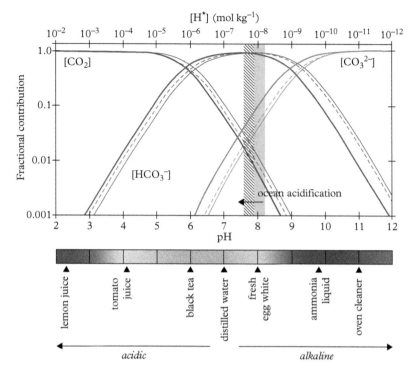

Figure 9.3 *Changes in dissolved carbon dioxide, carbonate and bicarbonate in the ocean, as a function of pH. The arrow indicates the direction of change under ocean acidification. The three curves indicate cases with variability for temperature and pressure. (After Barker & Ridgwell, 2012)*

the uptake of carbon dioxide in the ocean is not infinite and is limited by the amount of dissolved organic carbon (eqns (9.1a, b)). The Revelle factor determines the ability of the ocean to take up carbon dioxide and varies from 9 in low-latitude waters to 15 in the Southern Ocean at high latitude (Egleston et al., 2010). This buffering implies that, on average, the ocean takes up less than 10% of what one may intuitively think from a change in atmospheric carbon dioxide. In fact, this is one of the main causes why it may take tens to hundreds of thousands of years for the ocean to come back into equilibrium with changed carbon dioxide concentrations in the atmosphere.

The carbonate system we have described so far would provide a steady state insight into the ocean carbon cycle at timescales less than a million years long. However, we have omitted one important component that plays a key role in removing carbon from the ocean at million-year timescales: the production of marine biogenic carbonates by coralline algae (magnesium calcite), corals and pteropods (aragonite), and coccolithophores and planktonic foraminifera (calcite). Figure 9.4 shows a picture of the minute calcite shells of the modern coccolithophore species *Emiliania huxleyi*. The beautifully patterned plates at the outside of the cell are composed of calcite. Coccolithophores live

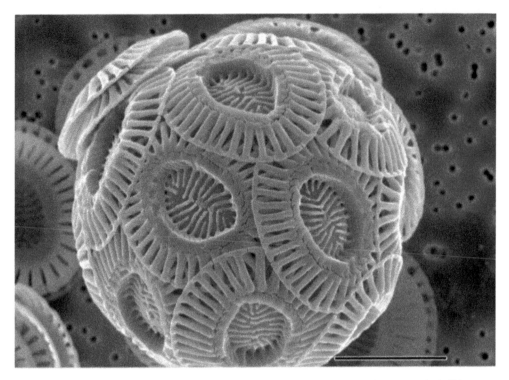

Figure 9.4 *Scanning electron micrograph of a single coccolithophore, Emiliania huxleyi. (Photo credit: Alison R. Taylor, Richard M. Dillaman, Bioimaging Facility, University of North Carolina, Wilmington)*

in the top layers of the ocean but, when they die, they sink, thereby transporting carbon into the ocean sediment. This carbon can remain there for many millions of years until taken up by the tectonic cycle.

The chemical reaction responsible for calcium carbonate (calcite) formation is

$$2HCO_3^{2-} + Ca^{2+} \rightleftarrows CaCO_3 + CO_2 + H_2O \tag{9.4}$$

While this process appears to produce one mole of carbon dioxide for each mole of calcium carbonate produced, the buffering of the ocean (e.g. eqns (9.1a, b)) works through changes in total alkalinity and TCO_2 in such a way that, under typical conditions, this is closer to a few tenths of a mole per mole of calcium carbonate (Zeebe, 2012). The burial of calcium carbonate in marine sediments represents an important pathway for removing carbon from the ocean–atmosphere system at million-year timescales, while the calcite shells provide important information on the ocean water conditions in which they were formed.

An important parameter in carbonate production and its subsequent precipitation is the calcium carbonate saturation:

$$\Omega = \frac{[Ca^{2+}][CO_3^{2-}]}{K_{sp}^{\star}} \tag{9.5}$$

where $K_{sp}^{\star}$ is called the solubility product of calcium carbonate, for either its calcite form or its aragonite form (two mineral forms of calcium carbonate, differing in their crystal forms). If $\Omega > 1$, we have saturation, and calcite can precipitate; if $\Omega < 1$, then we have undersaturation, and calcite dissolves. Since $K_{sp}^{\star}$ varies with pressure and temperature, there is a change in saturation with depth in the ocean. The depth at which the saturation ratio equals 1 is called the 'saturation horizon'. If changes in the alkalinity of the ocean occur, they will affect the saturation ratio.

9.3 The biological carbon pump

Coccolithophores and other plankton groups, such as diatoms, contribute to algal blooms in spring. As such, they are the visible expression of what is generally known as the biological pump (Figure 9.5). The carbonate pump we discussed in Section 9.2 is the other key mechanism for carbon dioxide uptake and outgassing in the ocean. Together with deep ocean transport (Chapter 7), they form the principal mechanisms by which oceans control the carbon cycle. When algae take up carbon dioxide, they convert it, under the influence of light into biomass and oxygen (see eqn (1.1)). This process depends critically on the availability of chlorophyll and key nutrients such as phosphate, nitrate and, for siliceous organisms, silicate, as well as iron. In fact, the light-capturing capabilities of chlorophyll allow detection of large algal blooms from satellite remote sensing. The autotrophic production of sugars and biomass is called Gross Primary Production but, because the building of biomass requires energy, part of Gross Primary Production is used for supplying this through respiration or burning of these sugars. Subtracting respiration from the Gross Primary Production yields the Net Primary Production, which is estimated to be 50 Pg C yr^{-1} in the global ocean. When biomass dies off, it is grazed on by heterotrophs (organisms that do not produce their own sugars). These consist of either zooplankton or bacteria (see Figure 9.5) and convert the dead biomass into dissolved organic matter or particulate organic matter. The dissolved organic carbon is partly converted into dissolved inorganic carbon. Only a fraction of the particulate organic matter eventually makes it to the ocean bottom. This flux is so small because of the effective recycling through zooplankton and bacteria in the upper layers of the ocean (0.05 Pg C yr^{-1} out of 4 Pg C yr^{-1} produced, i.e. less than 1%). In contrast, 25% of the calcium carbonate formed in the surface ocean makes it to the ocean floor (0.25 Pg C yr^{-1} out of 1 Pg C yr^{-1}). Note that, in this process, other components of biomass, such as nitrogen and phosphorus, are also recycled. The mean residence time of organic carbon in the ocean's surface is thus very small, of the order of days to weeks.

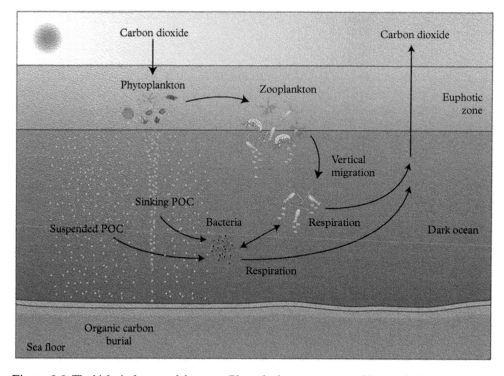

Figure 9.5 *The biological pump of the ocean. Phytoplankton are consumed by zooplankton or respired by bacteria and dissolved to organic carbon . The biological pump ultimately transports carbon to the ocean floor; POC, particulate organic carbon. Further explanation in the text. (After Herndl & Reinthaler, 2013)*

The ratio of the elements carbon, nitrogen and phosphorus is remarkably constant in marine algae and is known after its discoverer as the Redfield ratio: 106 C:16 N:1 P. In particular, the ratio of nitrogen to phosphorus is very stable. In cases of very high productivity, this implies that nitrogen and phosphorus very quickly become limiting and only a limited amount of inorganic carbon can be used for photosynthesis. Rivers, atmospheric deposition and upwelling currently produce only about 10% of the supply of nitrogen and phosphorus, indicating the necessity for rapid recycling of organic material in the upper layers of the ocean (i.e. the process shown in Figure 9.5). However, this percentage has not been constant in the geological past.

9.4 Ocean–air fluxes

Thus far, we have a biological mechanism and a physical–chemical pump that can take up, release and transport carbon dioxide. It is now important to ask the next question: to what gradients, concentrations, carbon stocks and fluxes do these pumps lead? Let us start with the flux (F) from the ocean to the air above. This is generally expressed by the diffusion equation

$$F = kK_0[(\mathrm{pCO_2})_\mathrm{w} - (\mathrm{pCO_2})_\mathrm{a}] \qquad (9.6)$$

where k is the transfer velocity, K_0 is the solubility, and $\mathrm{pCO_2}$ is the partial pressure of carbon dioxide at the water surface (subscript 'w') and in the air (subscript 'a'). The partial pressures can be measured with the devices that we discussed in Chapter 3. For the carbon dioxide at the water surface, the partial pressure is obtained after the water that has been sampled has reached equilibrium within a so-called head space on the instrument. After that, a gas analyser can determine the concentration. Alternatively, as often in the past, the partial pressure of carbon dioxide in water can be calculated from alkalinity and dissolved inorganic carbon (eqns (9.1a, b) and (9.2)).

The quantification of the transfer velocity presents other problems. Over a flat, solid surface, this would depend strictly on the physical properties of that surface and the wind speed, with the former unchanging. Over the ocean, or any other water surface, wave action complicates this simple relation dramatically. Various attempts have been made to account for the variability of surface roughness caused by waves (Wanninkhof, 2014). These range from assuming a flat surface, where winds are less important than other factors such as thermal gradients, to a situation where waves are caused by high winds, in which case transport through bubbles may further complicate matters. Generally, a cubic equation is used to correlate wind speed and transfer velocity over water, not unlike the evaporation equations used for conditions over land.

The value for $(\mathrm{pCO_2})_\mathrm{a}$ in eqn (9.6) is well known over today's ocean, being obtained from about eighty flask stations around the world (Wanninkhof et al., 2013). It varies seasonally by about 10 ppm over the Northern Hemisphere, driven by changes in photosynthesis and respiration over land, but much less over the Southern Hemisphere, where there is less land. This small variability is largely due to the very efficient mixing in the atmosphere. If we take this argument to the limit, one single station could well be representative of the global increase and variability of carbon dioxide. In fact, this is the reason behind the success of the Mauna Loa series shown in Chapter 2.

The $(\mathrm{pCO_2})_\mathrm{w}$ over the ocean is much more variable, of the order of 100 ppm seasonally and interannually and can be even higher in algal blooms. The spatial variability is also much larger. The correlation length scale for this variable (a measure of how correlated a variable is at one location to its value at another location) in the atmosphere is close to 1000 km; in the ocean, this is an order of magnitude less. This implies that one needs many more observations to determine this part of eqn (9.6). To enable us to calculate the global and regional sea–air fluxes, various interpolation techniques are used to compensate for the lack of observations. This adds considerable uncertainty to the estimates of carbon uptake and outgassing.

These calculations of the fluxes are sometimes calibrated against ^{14}C depletion rates in the atmosphere and increases in the oceans. The atomic bomb explosions in the 1960s caused the ^{14}C concentration in the troposphere to almost double. This has provided geochemists with an excellent tracer for carbon uptake. It is possible to determine the

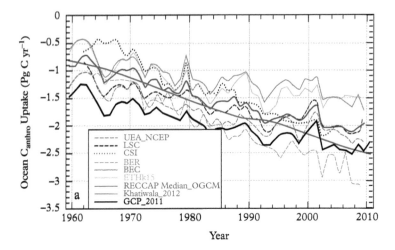

Figure 9.6 *Fifty-year globally integrated ocean anthropogenic carbon dioxide uptake from ocean biogeochemistry general circulation models (OBGCMs) used in the Regional Carbon Cycle Assessment and Processes project. The thin solid and dashed lines show the annual uptake of the different models and their interannual variability. The thick solid blue line is the median of the OBGCMs; the thick solid red line is the output of an inverse method (Khatiwala et al., 2013) and the thick black line is the result from a slightly different ocean model ensemble from the Global Carbon Project. (After Wanninkhof et al., 2013)*

transfer velocity of $^{14}CO_2$ from the oceanic uptake of bomb ^{14}C by the increase in ^{14}C in the ocean and its decrease in the troposphere. This scales with the transfer velocity of carbon dioxide. Taking into account improvements in the ^{14}C inventory of the ocean and measured wind speeds, including measurements from satellites, it is thus possible to calculate the ocean–air flux of carbon dioxide. The results of such calculations are shown in Figure 9.6 and suggest a slowly increasing ocean sink over the last fifty years, with the current uptake of about 2.5 Pg C yr^{-1}.

9.5 Ocean carbon stocks

Aside from the deep Earth reservoirs of carbon, the ocean carbon stocks are the largest in System Earth (see Figure 9.2). The surface ocean contains 900 Pg C, while the intermediate and deep sea contains the very large amount of 37,100 Pg C. Surface sediments in the ocean floor contain an additional 1750 Pg C. The amount of biomass in the ocean is small—only 3 Pg C—compared to the two orders of magnitude more in vegetation on land. Carbon dioxide dissolved in the seawater as dissolved organic carbon adds another 700 Pg C. On top of this, an estimated 155 (± 30) Pg C are added as a result of emissions of anthropogenic carbon dioxide due to fossil fuel burning emissions and from deforestation.

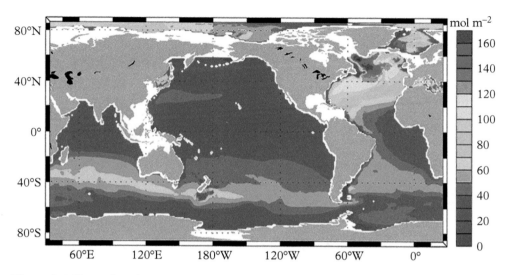

Figure 9.7 *'Best estimate' ocean anthropogenic carbon uptake. (After Khatiwala et al., 2013)*

Ocean stocks of anthropogenic carbon are hard to determine but, by using a combination of methods, including repeated sampling, inverse modelling and the use of human-made tracers (chlorofluorocarbons, bomb ^{14}C), it is possible to estimate the uptake of anthropogenic carbon by the oceans. The most recent such estimate (Khatiwala et al., 2013) gives the total anthropogenic carbon ocean uptake since the industrial revolution (1850) as 155 Pg C. The distribution of this uptake is spatially very uneven (Figure 9.7), with most of the uptake of anthropogenic carbon taking place in the Atlantic. This uptake is thought to be caused mainly by the physical pump and deepwater formation: as surface waters cool on their northward migration, they start sinking to larger depths, bringing the carbon with them to larger depths (see Chapter 7).

9.6 Terrestrial carbon

Biological processes largely drive the terrestrial carbon cycle; this is in contrast to the ocean carbon cycle, where the physical and chemical solubility pumps also play an important role. It is important to appreciate that each of the key processes in the terrestrial cycle has a different timescale associated with it and that most of these timescales are less than centuries long (Figure 9.8). This makes the terrestrial carbon cycle an important player in the short-to-medium term, rather than at geological timescales.

When vegetation absorbs solar radiation, and sufficient amounts of chlorophyll and nutrients are available, photosynthesis produces the sugars or energy required for growth

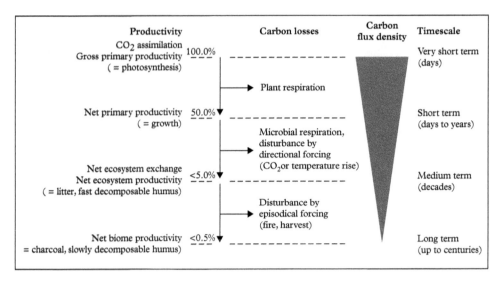

Figure 9.8 *Terrestrial carbon cycle definitions of Gross Primary Production, Net Primary Production, Net Ecosystem Production and Net Biome Production. The carbon ecosystem losses through respiration, decomposition and disturbance are shown as sideways arrows. The width of the big downwards-pointing arrow indicates the relative magnitude of the carbon flux. Timescales from days to centuries are also indicated.*

and maintenance. This process is called Gross Primary Production and is similar to the Gross Primary Production of the algae discussed previously. Current best estimates put a likely value of terrestrial Gross Primary Production at around 120 Pg C yr^{-1}, but its precise determination is difficult. Plants respire roughly 50% of this Gross Primary Production, leaving Net Primary Production at around 60 Pg C yr^{-1}. This is what we can consider to be the short-term carbon uptake. Note that this is roughly the same size as the ocean uptake (pre-industrial at 50–60 Pg C yr^{-1}). However, in the longer term (years to decades), plants lose their leaves or die off and fungi and other heterotrophic organisms decompose the material that reaches the soil (see Chapter 10 for a more elaborate discussion of respiration). As in the ocean, heterotrophic organisms depend on other organisms for their energy source, whereas autotrophic organisms can produce their own energy. The resulting Net Ecosystem Production reads in equation form as

$$NEP = NPP - R_{auto} - R_{hetero} \qquad (9.7)$$

where *NEP* is the Net Ecosystem Production, *NPP* is the Net Primary Production, and R_{auto} and R_{hetero} are the autotrophic respiration and heterotrophic respiration, respectively. Net Ecosystem Production is only a small fraction (generally <5%) of Gross Primary Production, so of the order of 10 Pg C yr^{-1}. Be aware that large regional differences can exist. For instance, in cold climates, heterotrophic respiration generally cannot keep up with Gross Primary Production, so long-term carbon sequestration in the form of peat or soil carbon may occur. In contrast, in high-temperature environments such as in the tropics, decomposition is fast and little soil carbon may be formed.

Over the longer term, periodic disturbances such as fires, or human management in the form of harvest of tree biomass, will reduce the Net Ecosystem Production further to a value of less than 0.5% of Gross Primary Production. This long-term storage of carbon is called Net Biome Production and is what really matters when we discuss the role of forest in mitigating climate change. Unfortunately, only too often in popular discussions, just Net Primary Production or even Gross Primary Production is considered, generating wildly exaggerated claims about the carbon-sequestration potential of our forests (see also chapter 13).

To illustrate the variability in carbon uptake and loss in natural forest, consider Figure 9.9. This shows the typical life sequence of a pine forest in central Siberia (near the Yenisei River). Here, the natural cycle contains fires, which hit these forests with a typical return period of about seventy years or so. When such a fire hits a stand, it will burn off most of the available carbon in a very short burst. This phenomenon, in which long-term slow carbon uptake and accumulation is lost abruptly, is typical of the terrestrial carbon cycle and is known by its maxim, 'slow in, fast out'.

When a forest starts regrowing in the first twenty to fifty years after such a fire, the whole ecosystem still loses carbon (negative slope), because the decomposition of

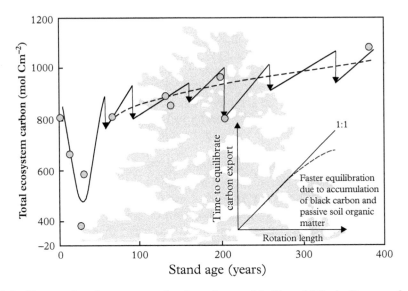

Figure 9.9 *Changes of total ecosystem carbon in a pine stand in Central Siberia. Downward arrows indicate carbon loss. The slope of the dashed line indicates the short-term Net Biome Production. (After Schulze et al., 2000)*

dead organic material provides a larger flux than that which is coming in through Gross Primary Production and Net Primary Production. It is thus fundamentally important to consider the full ecosystem—not just the trees or the above-ground biomass but also the below-ground biomass, dead wood and carbon in the soil. When the carbon dioxide losses of these components decline, the forest starts taking up carbon, and the slope of the line becomes positive again. Now, typical surface fires may hit the ecosystem and cause the loss of some carbon occasionally but, overall, the forest appears to work towards a steady state. However, and importantly, this steady state is never reached, as disturbances will continue to bring down the amount of carbon. So, old-growth forests can still take up carbon, and the role of disturbance is critical in their life cycle.

9.7 Terrestrial carbon fluxes

Changes in the stocks of carbon are the time integral of the fluxes of carbon, primarily Gross Primary Production, respiration and disturbance. There are several ways to estimate or observe fluxes. The long-term flux can be observed by taking the difference of two stock estimates divided by the time. This can be achieved by making repeated visits (e.g. every ten years) to the same site. The FAO forest inventory plots (see http://www.fao.org/sustainable-forest-management/toolbox/modules/forest-inventory/basic-knowledge/en/) provide such data.

Micrometeorological techniques, such as eddy covariance systems, that directly observe the Net Ecosystem Exchange can also be used. The advantage of these latter measurements is that they provide direct process knowledge of the controls and regulation of the fluxes. For instance, when measurements are made over more than ten years, the effects of episodic drought (or flooding) can be disentangled through process studies. The disadvantages are the costs, the fact that the system requires sophisticated instrumentation set up on a scaffolding or large tower so that the flux can be measured above the canopy, and the limited sample size (or flux footprint, the size of the area of vegetation that is measured and is representative for the flux). To get global coverage, a very large number of sites need to be set up. Currently, about 400 such sites exist (Baldocchi, 2014).

The other ways to estimate fluxes rely on models. Dynamic Global Vegetation Models can simulate the interannual variation of fluxes. Although they tend to agree when run for the historical period, they start to diverge widely in their response when used for future climate (Sitch et al., 2008). In general, however, they do appear to agree on the interannual variability, even if they disagree on the mean flux. The final method for inferring fluxes is to back calculate the flux from observed spatial carbon dioxide concentration gradients in the atmosphere. These gradients are observed using a network of stations (see Chapter 3). These inverse modelling techniques use an a priori estimate of the terrestrial flux and then optimize this according to the concentration gradients in the atmosphere. The a priori fluxes are often taken from Dynamic Global Vegetation Models. Figure 9.10 shows the results of several of these methods, including estimates of land-use change emissions for 1960 to 2015. Note that the amount of land-use change

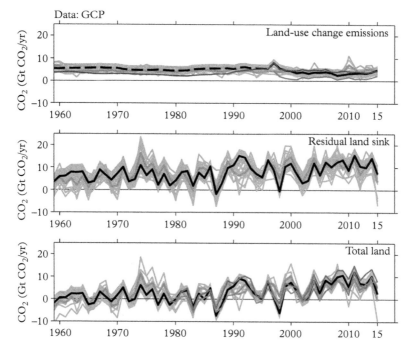

Figure 9.10 *Terrestrial carbon fluxes. The top panel shows the emissions from land-use change (mainly deforestation). The middle panel shows the land sink calculated as the residual of the atmosphere and ocean sink and anthropogenic emissions. The bottom panel shows the calculated land flux. The estimates from a series of individual land models (Dynamic Global Vegetation Models) are shown in green, Le Quéré et al's (2018) estimate is shown in black, and the inversion model estimates are shown in purple and red. (After Le Quéré et al., 2018)*

emission slowly decreases, and the land sink gradually increases. This increase is an important issue in the prediction of the future carbon budget, one which we will return to later in this chapter.

9.8 Terrestrial carbon stocks

Globally, the amount of carbon stocked in vegetation is a key variable in the carbon cycle—one that is very important but difficult to estimate. The classic way to estimate the amount of carbon in vegetation is to destructively sample it and then determine the amount of carbon in the dry weight. In forests this quickly becomes extremely difficult, and general (allometric) relationships between stem diameter and carbon content are used to get an estimate of the above-ground carbon (ABC) of vegetation. A serious disadvantage is that the measurements are rather location specific, so one needs either a very large number of sites or some clever upscaling technique to extrapolate them.

Remote sensing offers an alternative but, unfortunately, carbon content is not a variable that can be directly inferred from the radiative measurements made on board a satellite. Both the remote sensing and the in situ techniques have the disadvantage that the amount of carbon stored below ground remains out of sight and difficult to estimate.

At present, two techniques exist that allow us to use remote sensing estimates for biomass: a lidar technique that measures tree height, which is subsequently converted into carbon content, and a microwave measurement that estimates the amount of water in biomass and thus, with adequate calibration, can provide estimates of the global carbon content distribution. An additional advantage of the latter technique is that a relatively long record exists from satellites, so trends can be estimated as well. The global biomass distribution from this technique (called vegetation optical density) is shown in the left panel of Figure 9.11. Clearly, most of the world's ABC is stored in tropical forests, with boreal forests and tropical savannas in second and third place, respectively. Note also the amount of ABC in crops, which is an important variable to include in the global carbon cycle, as roughly 50% of Earth's surface is under some form of agriculture. Total above-ground biomass is estimated to have been 362 Pg C in 2000, although the uncertainty is large: the estimate is bounded by 90% errors of 310–422 Pg C. Note that this is still ABC only. Liu et al. (2015) used biome-specific conversion factors to convert this into total carbon content. These factors convert ABC into total carbon by multiplying ABC, and are largest for boreal forest, at 5.2, while, for temperate and tropical forests, these are 3.2 and 1.8, respectively. This reflects the observations that boreal forests store a much larger proportion of carbon below ground than do tropical forests. We saw this before when we discussed Net Ecosystem Production in Section 9.6. Deforestation in the humid tropical forests is still the main loss term in ABC, while substantial greening (biomass expansion) is taking place in the boreal forest.

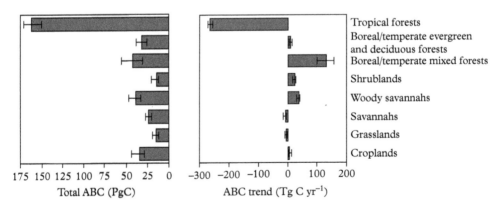

Figure 9.11 *(Left) Global stocks of above-ground carbon (ABC), as estimated from satellite observation of vegetation optical density, in pentagrams of carbon. (Right) Trends in carbon stocks from 1993–2015, given in teragrams of carbon per year. (After Liu et al., 2015)*

9.9 Mean residence time of carbon on land

Stocks and fluxes combine in mean residence or turnover time (see Chapter 1). This important carbon cycle variable defines the average time a carbon molecule stays in a particular reservoir. Figure 9.12 shows the turnover time for the carbon in vegetation and soil in more detail; for this figure, Carvalhais et al. (2014) used satellite estimates of the stock of carbon and scaled global estimates of the flux from eddy covariance systems to produce a map of the turnover time of carbon. Clearly, large variations exist but, on average, carbon resides in the vegetation and soil near the equator for a substantially shorter time than it does at latitudes north of 75° N (mean turnover times of 15 and 255 years, respectively). Compare this to the rapid turnover in the surface ocean! The longest turnover times are found in the cold biomes (tundra, 65 years; boreal forests, 53 years), in temperate grassland and shrubland (41 years) and in desert areas (36 years). As expected, tropical forests and savannas have the shortest turnover times (14 and 16 years, respectively). These estimates are based on Net Ecosystem Production, not Net Biome Production, and carry significant uncertainty. A logical extension to this work would involve making maps of Net Biome Production, but one can surmise that the net effect would be to increase the mean residence time if the same carbon stock were divided by a smaller flux.

Correlating these turnover times with precipitation and temperature, Carvalhais et al. (2014) found that temperature first and then precipitation played the most significant roles in determining the turnover time. Warm temperature affects growth and thus the size of the flux, while cold temperature restricts not only growth but also respiration, as we have seen before. Availability of water, or rather the absence of it, affects growth negatively, and this is visible in the areas that are prone to high variability in rainfall, such

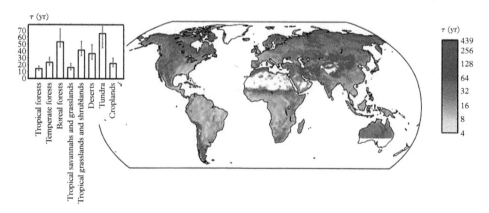

Figure 9.12 *Global distribution of total ecosystem carbon turnover time τ. The left-hand inset gives turnover times of carbon in terrestrial ecosystems. (After Carvalhais et al., 2014)*

as the semi-arid and Mediterranean regions. The authors also noted that current Earth system models have serious difficulties reproducing these observational correlations and, consequently, underestimate the role of hydrology in the carbon cycle.

9.10 Geological carbon cycle

Let us now look at the deep-time carbon cycle (timescale $>10^6$ years) with some of the measurement tools, models and processes we have described earlier. Carbon can come in either organic or inorganic forms (see Chapter 1). The sum of all carbon always has to be constant when outgassing of carbon dioxide through the mantle or volcanoes, and other sources of carbon, and the burial, carbonate and organic carbon do not change too much over time. Note again that, of all reservoirs of carbon, Earth's mantle contains by far the largest amount ($\sim 60 \times 10^6$ Pg C). The ratio between organic and total carbon, f_o, is an important parameter and can be determined from the stable isotope ratios of carbon, as they have to follow the same mass balance as the total carbon does (Langmuir & Broecker, 2012). Some algebraic manipulation then gives their ratio as

$$f_o = \frac{\delta^{13}_{ic} - \delta^{13}_{tc}}{\delta^{13}_{ic} - \delta^{13}_{oc}} \qquad (9.8)$$

where the subscripts denote inorganic (ic), organic (oc) and total carbon (tc). If we take these measurements at rocks from different ages, we can estimate how this fraction varies through geological time. Importantly, the ratio is equivalent to what is called the burial rate, the amount of reduced organic carbon that is not available for oxidation.

The lower panel of Figure 9.13 shows $\delta^{13}C$ for carbonate (inorganic carbon) and organic fractions over geological time. There are two important things to note: (i) the offset between inorganic and organic carbon is generally 30‰ and (ii) mantle $\delta^{13}C$ is always −5‰. Therefore, if no organic carbon were produced, the value of $\delta^{13}C$ would thus be approximately −5‰. The first organic carbon produced would incorporate the offset of 30 ‰ and obtain a value of 35‰. According to Figure 9.13, the level of carbonate carbon has always been higher than that of the mantle by ~5‰. This implies that, to balance, the organic (non-mantle) $\delta^{13}C$ should be around −25‰ and that carbonate carbon makes up about 80% of Earth's active carbon. In other words, this suggests that the long-term average burial rate of organic carbon is around 0.2 (see the upper panel of Figure 9.13; also see eqn (9.4)). The long-term variations in $\delta^{13}C$ values over time also suggest that, from the Archean (4 Gyr ago) to the present, there has been an increase in the burial rate of organic material by about 15%–25%.

Interestingly, the $\delta^{13}C$ values show some significant excursions around 2.0–2.4 Gyr ago and 0.5–0.8 Gyr ago. These suggest higher proportions of organic matter and greater production of oxidizing power, which in turn suggests that these values may be related to Earth's oxygenation, which started at around 2.5 Gyr ago, with a subsequent surge of oxygen around 0.8 Gyr ago (see Chapter 1). The period from 1 to 2 Gyr ago is sometimes referred to as the 'boring billion', due to its constancy in $\delta^{13}C$ values and corresponding burial fractions.

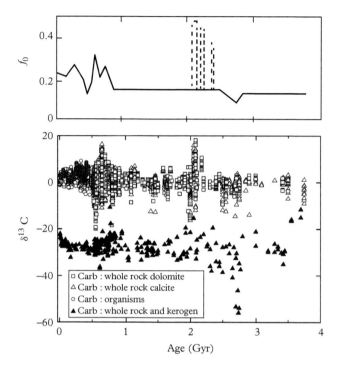

Figure 9.13 *Upper panel: schematized burial rate of organic carbon. Lower panel: $\delta^{13}C$ for carbonate and organic material through geological time; open squares, carbonate from whole rock dolomite; open triangles, carbonate from open rock calcite; open circles, carbonates from organisms; closed triangles, organic carbon from whole rock and kerogen. (After Wallmann & Aloisi, 2012)*

9.11 The Paleocene–Eocene Thermal Maximum

The period around 55 Myr ago also shows a strong negative excursion in $\delta^{13}C$ values. This is known as the Paleocene–Eocene Thermal Maximum. The period is of considerable interest because the values of $\delta^{18}O$ in benthic foraminifera suggest a sharp increase in temperature of about 5 °C over a relatively short period, less than 10 kyr. Earth was a very different place then, with very high carbon dioxide concentrations in the atmosphere (Figure 9.1) and no ice sheets near the poles. In fact, this period may be our best geological confirmation of the greenhouse gas theory. As said, the timescale at which this large $\delta^{13}C$ excursion took place is estimated to be around 10 kyr, while it took about ten times longer, that is, 100 kyr, to get the carbonate system and atmospheric carbon dioxide back to its initial state. The input of carbon was removed primarily through silicate weathering and precipitation of carbonate in the ocean, and carbon uptake by the biosphere and subsequent burial as organic carbon.

The timescales of change make the study of the Palaeocene–Eocene Thermal Maximum of considerable relevance to the current climate, where we also have a sharp increase in atmospheric carbon dioxide over a relatively short period, albeit in a couple hundred years rather than in thousands of years. We also would expect a sharp excursion towards negative $\delta^{13}C$ values, as carbon dioxide from the burning of fossil fuels is generally depleted in ^{13}C. Analysis of tree rings and sponges in the ocean has indeed confirmed this. As for the Palaeocene–Eocene Thermal Maximum, we also see an increase in acidity of the current ocean because of the additional ocean uptake of carbon dioxide. Models can be used to estimate the rate of carbon release and timing more precisely, with Zeebe et al. (2016) finding an estimated 0.6–1.1 Pg C yr^{-1} release rate over 4000 years. For comparison, our current (as of 2016) rate of carbon release into the atmosphere is close to 10 Pg C yr^{-1}. However, the carbon dioxide is supplied over a much shorter time period, and the input is unequivocally anthropogenically driven. Given our current understanding of the paleorecord, the current rate appears unprecedented over the last 60 Myr.

The question remains as to what caused this sudden negative carbon isotope excursion in the Palaeocene–Eocene Thermal Maximum. The carbon injected into the atmosphere–ocean system must have originated from a ^{13}C-depleted source, implying either organic matter or methane. The lighter the isotopic value of the source, the smaller the amount of carbon that must be released to explain the isotopic shift. It is still a question whether an earlier warming triggered the release of carbon or whether the carbon release itself caused all warming. For instance, a release of ~2000 Pg C from methane hydrates (see Chapter 10) could explain the excursion, but significant warming should then have preceded the carbon input to trigger the methane hydrate dissociation (see Chapter 10). In contrast, a release of at least twice this amount of carbon from peat or permafrost reservoirs must be invoked to explain the negative excursion by the oxidation of organic matter. Recently a source of 10,000 Pg C yr^{-1} was identified from an analysis of $\delta^{13}C$ and a pH reconstruction assimilated in an Earth system model (Gutjahr et al., 2017). The low $\delta^{13}C$ (−17‰) they established points to a source of volcanic origin associated with the opening of the North Atlantic.

So, although the $\delta^{13}C$ excursion is not disputed—on the contrary, the geological Paleocene–Eocene boundary is defined by a clay layer that corresponds to it—what caused it and, to some extent, what sustained the low $\delta^{13}C$ of the input into the atmosphere, are still open questions.

9.12 Carbon cycle in glacial–interglacial cycles

Let us now move somewhat closer to today in geological time (some 10^4–10^5 years). We saw in Chapter 2 that the concentration of atmospheric carbon dioxide during Pleistocene glacial–interglacial cycles varies from roughly 180 ppm to 280 ppm, with the low values occurring at glacial times (see Figure 2.3). What precisely causes this variability and what its main drivers are still very much open questions and is considered to constitute the 'holy grail' of carbon cycle science. Several drivers have been identified through modelling and data analysis since Wallace Broecker in 1982 first posited the important role of the ocean with its biological pump and calcium carbonate burial in pacing the glacial–interglacial cycles. The estimated impact on the atmospheric carbon dioxide concentration of several proposed drivers is shown in Figure 9.14, which is taken from the latest IPCC report (Ciais et al., 2013).

Let us start with the process that we have just described for the deep-time carbon cycle, terrestrial weathering. This is unlikely to have played a major role, given the timescale at which the processes of weathering operate; the fluxes involved are simply too small to generate the large swings in carbon dioxide concentrations observed from the ice cores (~90 ppm). The second terrestrial component, carbon storage on land, works generally in the wrong direction. In colder climates, land carbon storage would decrease rather than increase (despite some more regional increases such as in permafrost areas). The $\delta^{13}C$ record as determined from oceanic benthic foraminifera confirms this.

Moving to the role of the ocean, we can identify sea-surface temperature as an important driver. Figure 9.14 suggest that the increase of sea-surface temperature during, for instance, the Last Glacial Maximum, of about 3–5 °C could have caused an increase in atmospheric carbon dioxide of roughly 25 ppm. Increasing solubility of carbon dioxide at low temperatures is the dominant cause (see Section 9.2), but this estimate contains a large uncertainty due to the difficulty of precisely determining the variation in sea-surface temperatures in the past. A sea level reduction of >100 m due to the water stored in the large ice caps would also have increased the salinity and alkalinity of the ocean waters, thereby increasing the atmospheric levels by a further 15 ppm.

So far, none of these processes is itself sufficient to generate the large >90 ppm glacial to interglacial swing observed in the ice cores. The role of the deep ocean has also received much attention. This should not come as a surprise, because the deep ocean stores by far the largest amount of carbon on Earth. Prolonged and significant changes in the deep ocean circulation or composition certainly would have the potential to change

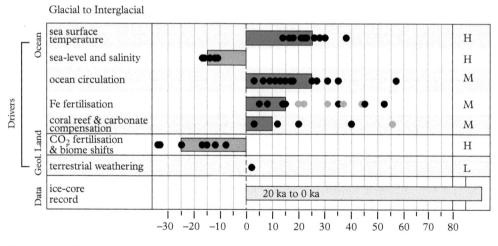

Figure 9.14 *Possible drivers of glacial–interglacial transitions, and their potential effects on atmospheric carbon concentration (in parts per million). (After Ciais et al., 2013)*

the atmospheric carbon dioxide concentration over glacial to interglacial timescales. That the ocean plays such a role is suggested by observed tight coupling of deep ocean temperature and atmospheric carbon dioxide concentration. The mechanism that is most likely to be responsible for this is increased stratification of the Southern Ocean. This would be able to dramatically reduce the release of carbon dioxide and thus increase the storage of carbon in the ocean during glacials. Model estimates of the role of the deep ocean circulation vary from an increase of only 3 ppm to 57 ppm.

Let us explore the role of ocean circulation in combination with the biological pump and ocean biogeochemistry in a bit more detail—it presents a good example of how physics and biogeochemistry work together in the ocean to impact climate (e.g. Sigman & Boyle, 2000). We will now look at the changes in ocean circulation and biogeochemistry from an interglacial to a glacial (Figure 9.14 showed the changes from a glacial to an interglacial). The most obvious culprit for a significant role in the carbon cycle is the Southern Ocean, due to its sheer size. Figure 9.15(a) shows the circulation of the deep ocean under 'normal' conditions, in which case the circulation is, by and large, equivalent to the thermohaline circulation discussed in Chapter 7. We consider this situation representative of an interglacial. North Atlantic Deep Water is formed in the northern polar seas, while Antarctic Bottom Water is formed in the southern seas. At the surface, the westerly winds and the Coriolis force combine to generate the currents. Topography influences such features as the Indonesian Throughflow and the Agulhas eddies around the coast of South Africa, which transport surface water from the Pacific to the Atlantic. Antarctic Intermediate Water and Subantarctic Mode Water provide transport at intermediate levels. Also shown are the Antarctic Circumpolar Current, the Polar Antarctic Zone and the Subantarctic Zone.

This description is of the transport of water only, but the amount of carbon that is transported through these currents is equally important. When nutrients such as nitrogen, phosphorus and iron are available, phytoplankton will use these with sunlight to produce biomass. The net effect of this is to deplete the surface of nutrients, while the dead biomass starts to sink to the ocean floor. Here, decomposition frees the nutrients and releases carbon dioxide through respiration. The net effect is then a downward transport of carbon. The nutrients subsequently become available through upwelling at higher latitudes after having been picked up by the dense deep-water flows.

When all of the nutrients are 'regenerated', the biological pump is said to work to full efficiency. When not all of the nutrients are used, the nutrients are called 'preformed' and, in effect, represent a wasted opportunity for algae to sequester carbon dioxide. Worse, the carbon dioxide from the deeper layers that was originally sequestered by the regenerated nutrient flow is now released. The low-latitude oceans thus constitute a fully efficient biological pump (virtually all nutrients are regenerated), whereas the high-latitude oceans generate mostly preformed nutrients and are unable to use them all. The Southern Ocean in interglacial conditions is far from being an efficient pump. The Antarctic Circumpolar Current also upwells (preformed) nutrient-rich waters that contain carbon dioxide and which are subsequently blown towards the equator. This moves carbon dioxide from the high latitudes to the low latitudes and is known as the Southern Ocean Carbon Leak. This is the situation at interglacials or in the modern ocean.

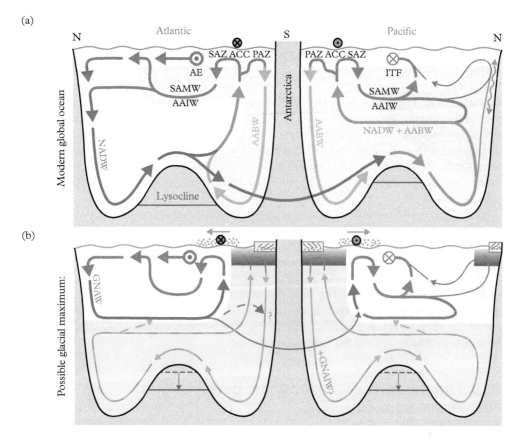

Figure 9.15 *Diagram detailing the possible changes in ocean circulation from (a) the current climate state to (b) a glacial state (Sigman & Boyle, 2000); AABW, Antarctic Bottom Water; AAIW, Antarctic Intermediate Water; ACC, Antarctic Circumpolar Current; AE, Agulhas eddies; GNAIW, Glacial North Atlantic Intermediate Water; ITF, Indonesian Through-Flow; NADW, North Atlantic Deep Water; PAZ, Polar Antarctic Zone; SAMW, Subantarctic Mode Water; SAZ, Subantarctic Zone (ITF and AE return surface water from the Pacific to the Atlantic). Circled points show transport out of the page, and crosses show transport into the page. The colour of the lines showing interior flows indicates their ventilation source region: blue, NADW or GNAIW; yellow, AABW; green, mixed NADW and AABW. Line thickness indicates changes in flow rate. Darker shading in the interior indicates a higher concentration of regenerated nutrient and excess carbon dioxide. (After Sigman et al., 2010)*

The hypothesis that now helps us find an explanation of the large swings in atmospheric carbon dioxide during interglacial–glacial transitions, involves stopping this Southern Ocean leak. This could be achieved by reducing the nutrient concentration of the region's input of new water, by reducing the deep-water volume that it ventilates or, more likely, a combination of these two mechanisms. Reducing the nutrient concentration occurs when productivity increases and more carbon dioxide is drawn into the ocean. Paleo-oceanographic data on productivity appear to confirm this scenario. The primary effect would then be strengthened by changes in the alkalinity system of the ocean: as more dissolved inorganic carbon remains in the oceans, they would become less acidic, allowing even more carbon dioxide to be drawn into the sea by the solubility pump. These hypothesized changes in circulation during a glacial state are shown in Figure 9.15(b). Here, essentially the ventilation of the deeper oceans becomes shallower, both in the Atlantic and in the Pacific, while the exchange at the Polar Antarctic Zone is shifted towards the equator. A freshwater lens and sea ice cover close to the pole further contribute to stemming the leak. Increased levels of dust containing iron needed for photosynthesis, derived from a glacial atmosphere, could then have provided further increases in the productivity (see Chapters 5 and 12), drawing additional carbon dioxide into the ocean.

   In summary, all these processes could have worked together to produce more storage of carbon in the ocean and less in the atmosphere during a glacial, as we can indeed observe in the ice cores. The correlation in timing of these changes in the carbon cycle and climate also holds clues for finding the more precise causes. However, increased time-resolution data show these processes to be far from gradual and indicate that processes at the centennial rather than the millennial timescale are also important. So, while the glacial–interglacial climate changes and the role of the carbon cycle in them are two of the most intriguing problems in the field of biogeochemistry and climate, these particular problems are still far from being solved.

9.13 The modern anthropogenic perturbation

Since the beginning of the Industrial Revolution around 1850 (we are now at the time-scale of hundreds of years), humans have relied on the burning of coal, oil and gas, commonly known as 'fossil fuels', for their energy supply. This use of fossil fuels has generated large emissions of carbon dioxide and other gases as a product of the burning of carbon (see eqn (1.1)). While there is some debate about whether humans produced any signs of their activities in the atmosphere before that date through agricultural activities (e.g. Ruddiman, 2005), the starting date of an exponential increase in modern day atmospheric carbon dioxide is unequivocal and undisputed. Figure 2.5, which shows the measurements taken at Mauna Loa since 1958, is the iconic representation of this increase.

   It is important to realize that this injection of carbon dioxide comes on top of what we have seen is an active carbon cycle with very large natural fluxes and fluctuations. It is thus not immediately obvious what effect the modern anthropogenic addition will have. Figure 9.16 shows the effects of the human perturbation of the fluxes in more detail.

   It is useful to remember that the carbon cycle, like any other cycle, runs with a closed budget, so the sum of the sources must equal the additional land uptake, the ocean uptake and the atmospheric increase, that is, any input must be met with a corresponding output:

$$F_{fos} + F_{landuse} = F_{land} + F_{ocean} + F_{atm} \qquad (9.9)$$

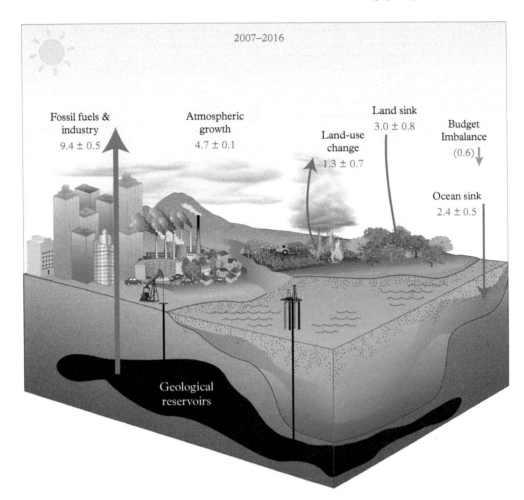

Figure 9.16 *Schematic representation of the overall perturbation of the global carbon cycle caused by anthropogenic activities, averaged globally for the decade 2006–15. The arrows represent emissions illustrated in eqn (9.9) from fossil fuels and industry (corresponding to F_{fos} in eqn (9.9)), emissions from deforestation and other land-use change ($F_{landuse}$), the growth rate in atmospheric carbon dioxide concentration (F_{atm}) and the uptake of carbon by the ocean (F_{ocean}) and land (F_{land}). All fluxes are shown in petagrams of carbon per year. (After Le Quéré et al., 2018)*

where F is the carbon dioxide flux, and the subscripts 'fos', 'landuse', 'land', 'ocean' and 'atm' indicate fluxes due to fossil fuel burning, land-use change emissions and the additional sinks of land, ocean and atmosphere, respectively. It is immediately obvious from Figure 9.17 that not all the carbon dioxide emissions from fossil fuel burning (and cement production) remain in the atmosphere. Over the last decade (2006–15), roughly 45% have stayed in the atmosphere, the remaining 55% being taken up by the land

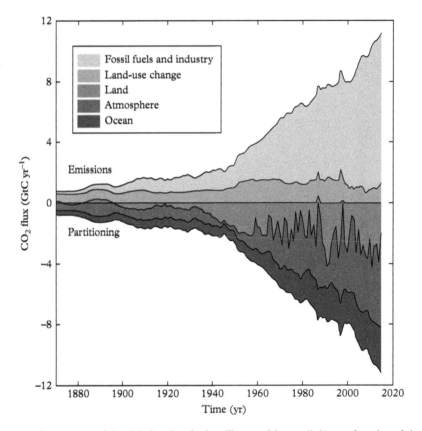

Figure 9.17 *Components of the global carbon budget illustrated in eqn (9.9) as a function of time, for emissions from fossil fuels and industry (corresponding to F_{fossil} in eqn (9.9)) and emissions from land-use change ($F_{landuse}$), as well as their partitioning among the atmosphere (F_{atm}), land (F_{land}) and oceans (F_{ocean}). (Le Quéré et al., 2018)*

(30%) and the ocean (25%). So far, this is good news. In the atmosphere, the radiative forcing is important; on land, the carbon dioxide can be usefully converted to biomass while, in the ocean, it can be absorbed. The ocean uptake, however, is not necessarily always a beneficial property, as uptake of carbon dioxide in water will inevitably lead to decreasing alkalinity, or ocean acidification.

Ocean acidification (see eqn (9.2)) is therefore referred to as the 'other carbon dioxide problem' and may have drastic consequences for changing the ecological boundaries of life in the ocean. Importantly, it will also affect the calcite or aragonite saturation and thus impact the building of calcite or aragonite shells. Through this process, it also has a direct impact on the long-term carbon cycle in the ocean. It is good to realize that the ocean and the land currently take up 55% of the emitted carbon. One of the big questions is for how long they will continue to do that. But, before we discuss that question, it is important to see how the budget was partitioned in the past. Figure 9.17 shows the partitioning in more detail. As mentioned before, the total sum of the emissions (sources) has to equal the total of the sinks. In this graph, the sinks are shown at the bottom, and the sources are given as positive fluxes at the top. We can see the exponential increase in fossil fuel emission and a somewhat slow decline in the land-use change emissions. The latter are primarily due to deforestation in the tropics. Note also that, up to around 1950, the emissions from land-use change were of the same size as those from fossil fuel burning. Up to that time, changing land use and burning of biomass were the greatest contributors to increases in atmospheric carbon dioxide. At present, human fossil fuel burning and cement production emit around 10 Pg C annually, while human emissions due to land-use change appear to be stabilizing at around 1.0–1.5 Pg C annually.

On the sink side, we see a rather constant oceanic uptake, increasing gradually to about 2.5 Pg C yr^{-1}. The atmospheric increase is much more variable, as is the land sink. At this timescale, it is the land sink that very much determines the variability in atmospheric carbon dioxide. In particular, El Niño episodes or large volcanic explosions have drastic impacts on the land uptake, which shows repeated minima around these years. In this graph, the most robust observations are those of the atmospheric increase. These are based on the solid measurement methods described in Chapter 3. The ocean uptake is modelled and the land uptake then appears as a residual. The remaining uncertainty for the land flux is estimated to be around ±0.8 Pg C yr^{-1} (Le Quéré et al., 2017). So, overall, we see land and ocean sinks increasing at the same pace as the increase in emissions due to fossil fuels. As a consequence, the ratio of atmospheric increase in carbon to the human emissions, called the atmospheric fraction, remains fairly constant at 45%. Particularly for the land uptake, there is no fundamental reason why this should be the case and also, importantly, no reason why this should continue in the future. The atmospheric fraction, therefore, is the parameter to watch: if it increases, it implies that the sinks cannot keep pace with our emissions. The big questions in today's carbon cycle are thus what causes the land and ocean to keep taking up carbon at increasing rates, are there any regional differences and sensitivities in this uptake, what causes its interannual variability and, importantly, when will they saturate?

9.14 The future carbon cycle

To answer these questions, we need to understand the key sensitivities of the land and ocean to increasing carbon dioxide levels and to a changing climate. Increasing carbon dioxide will provide more resources to plants for photosynthesis. The first-order effect of carbon dioxide fertilization is thus an increase in Gross Primary Production, particularly on land but also in the ocean. In the longer term, this has to be balanced by a similar increase in respiration (or burial, at geological timescales) as more material becomes available for decomposition. It is also not clear what an increase in Gross Primary Production does further down the carbon chain on land (see Figure 9.8): will it increase Net Biome Production by a similar amount, or is the carbon 'lost' on the way? A further critical issue in estimating the carbon dioxide fertilization effect is the limitation of nutrients; while plants initially will benefit from the increased resource available, at some time, the nitrogen or phosphorus available may not be enough to sustain this increase, like our discussion of the Redfield ratio in the ocean. Experiments lasting several years with artificially carbon dioxide-enriched air seem to point in this direction. In a similar way, for the ocean, increased atmospheric carbon dioxide will increase ocean uptake but, if decreasing alkalinity becomes such that this is no longer possible, the uptake will falter. It comes at a price.

The response of the ocean and land to increased temperature and changing precipitation is thus key to understanding the future carbon balance. Generally, increased temperature will increase carbon dioxide uptake on land, but it will also increase respiration. The expectation is that, at some time, maybe even in this century, this will lead to an imbalance between uptake and respiration, with the land turning from a sink to a source. To be able to predict this adequately, one needs coupled models including biogeochemistry, of the ocean, land and atmosphere. Their simulated response to changes in radiative forcing should include changes on land, where good plant physiology provides a realistic response to carbon dioxide fertilization, and in the ocean, where an adequate carbon cycle takes into account changes in solubility, transport and the biological pump. The frank assessment at present is that such good models do not yet exist. However, it is possible to investigate the order of magnitude of the response using the current state-of-the-art models. This has been the case in the coupled carbon cycle climate–model intercomparison project (C4MIP) and CMIP5, under the auspices of the World Climate Research Programme.

These models are used to analyse the future response of the carbon cycle to changes in climate and carbon dioxide from the perspective of two opposing feedbacks: the concentration–carbon response (β) that determines changes in storage due to elevated carbon dioxide, and the climate–carbon response (γ) that determines changes in carbon storage due to changes in climate. The results show (e.g. Ciais et al., 2013) that β is always positive, both for land and for ocean, and that γ tends to be negative.

In Figure 9.18, the regional differences in these responses are plotted for a number of the models. Clearly, the dominant effect of the concentration response is expected to be in the tropics, where the land shows a strong response to elevated carbon dioxide. The response in the ocean is less strong and is primarily located in the North Atlantic and the Southern Ocean. Models that include a nitrogen component generally show a much reduced response, as nitrogen availability limits further future increases in carbon storage. The climate–carbon response is generally negative, except for the boreal regions, where the previous temperature limitation on plant growth is removed by a warming climate. Again, the largest effect is in the tropics, where increased respiration and changing vegetation patterns decrease the potential for carbon storage. The uncertainties are, however, very large, as the shaded areas in the right-hand side plots in Figure 9.18 show.

In closing this chapter, it is illuminating to paraphrase some of the IPCC conclusions in the 2013 assessment report (Ciais et al., 2013):

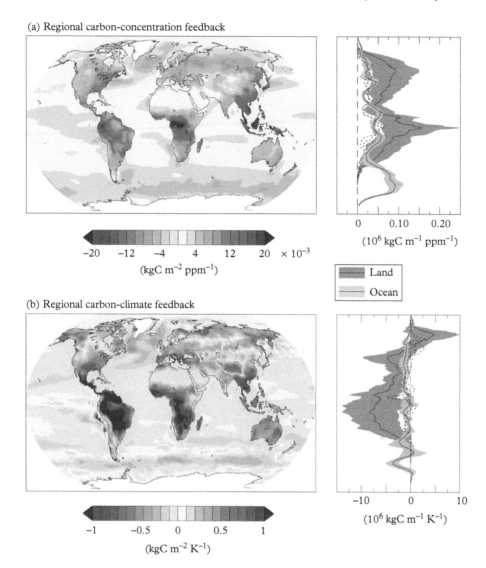

Figure 9.18   *(a) The regional distribution of the concentration–carbon response (β) and (b) the climate–carbon response (γ). The latitudinal plots on the right-hand side show the shaded area as the 1-sigma uncertainty. The dashed lines show results of land models that include a nitrogen component that results in considerable less response in both feedbacks. (After Ciais et al., 2013)*

- during the last 7000 years, the atmospheric carbon dioxide concentration varied between 260 and 280 ppm but it is currently (as of 2018) just above 400 ppm;
- during the past 900,000 years, the atmospheric carbon dioxide concentration varied from 180 ppm in glacials to no more then 300 ppm in interglacials;
- the ocean carbon uptake is likely to continue up to 2100, but land may switch from being a carbon sink to a carbon source;
- climate change may partly offset the increased sinks due to elevated carbon dioxide.

The carbon–climate feedback has implications for the amount of fossil fuel humans can use before a threshold in temperature is reached. If the feedback is overall positive, that is, more carbon dioxide comes into the atmosphere due to the feedback, the amount of usable fossil fuel reserves becomes proportionally smaller. We will come back to this aspect in the final chapters.

10

Methane Cycling and Climate

10.1 Methane

Methane is the most common reduced compound found in the atmosphere. It is produced mostly by microorganisms digesting organic material under anaerobic conditions. If you were to look for the presence of life on other planets, methane would be one of your first choices, as it signifies an atmosphere in disequilibrium. The disequilibrium on Earth is sustained by continuous production of methane by life. On Earth, only small amounts are produced by non-biological mechanisms, such as biomass burning and production deep in the ocean near thermal vents.

Methane is the second most important anthropogenic greenhouse gas. It strongly absorbs infrared radiation in two absorption bands, at wavelengths of 3.3×10^{-6} m and 7.7×10^{-6} m, which are related to the stretching and bending modes of the C–H bond (see Chapter 4). Although the concentration of methane is about 200 times lower than that of carbon dioxide, its global warming potential is 28 over a 100-year time horizon (Chapter 4). In today's atmosphere, carbon dioxide is the most abundant, at 400 ppm versus 1.8 ppm for methane.

Life produces waste. The rise of atmospheric methane is thus directly coupled to the occurrence of the first photosynthetic bacteria that produced organic waste in then still anoxic environments. There are two major pathways for producing methane:

$$CH_3COOH \rightarrow CO_2 + CH_4 \qquad (10.1a)$$

$$CO_2 + 4H_2 \rightarrow CH_4 + 2H_2O \qquad (10.1b)$$

The first, eqn (10.1a), is the acetate pathway, which is probably the older of the two and consists of splitting the acetate molecule into carbon dioxide and methane. The microbial organisms that perform these reactions are from the domain Archaea, single-cell prokaryotes which survive today in wetlands, in coastal sediments and near hot water vents. They are thought to have developed very early in the Archean, around 3.8 Gyr ago, just after the arrival of the first bacteria.

Biogeochemical Cycles and Climate. Han Dolman, Oxford University Press (2019). © Han Dolman.
DOI: 10.1093/oso/9780198779308.001.0001

10.2 The terrestrial methane budget

Where is methane being produced at present? It is worth distinguishing between natural and anthropogenic processes. Methane is produced in anoxic conditions in natural wetlands (hence its name 'marsh gas'), in agriculture from enteric fermentation by ruminants, and rice paddies and by biomass burning. Decomposition of waste in landfills is also a significant anthropogenic source, as is the methane production associated with coal, oil and gas extraction for fossil fuel, such as leaks and release of shale gas. Figure 10.1 shows the main source categories of methane and their ranges based on a recent global methane budget (Saunois et al., 2016). The largest natural source of methane is wetlands, while human sources from waste and agriculture almost equal this.

It is important to stress that not all methane that is produced in soils or wetlands enters the atmosphere. During its transport to the atmosphere, methane can also be oxidized to carbon dioxide by methanotrophic bacteria. This process makes modelling natural methane emissions difficult: not only do you need to know the extent of wetland, but you also need to know its production and consumption accurately. In contrast to inversions that estimate the source from atmospheric observations of methane concentration (Chapter 3), so-called bottom-up estimates of these emissions consequently vary considerably, from $155 \, \text{Tg CH}_4 \, \text{yr}^{-1}$ to $227 \, \text{Tg CH}_4 \, \text{yr}^{-1}$. The uncertainties in this budget are thus large—much larger than, for instance, some of the terms in the total carbon budget. Uncertainties are estimated to be 20%–30% for the inventories of anthropogenic emissions in each sector (agriculture, waste, fossil fuels) and for biomass burning. For natural

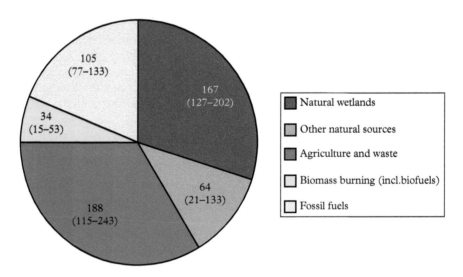

Figure 10.1 *The main source categories of methane emission to the atmosphere, based on the top-down budget for the period 2003–12; numbers are in teragrams of methane per year. Under natural sources, marine and freshwater comprise 68% of the total. (Values from Saunois et al., 2016; graphics, after Dean et al., 2018)*

sources, they are much larger: estimates vary by 50% for natural wetland emissions and by 100% or more for other natural sources (e.g. inland waters, geological sources).

The uncertainty in the dominant sink of global methane loss via hydroxyl radicals is estimated to be between 10% and 20% (Kirschke et al., 2013). On this sink side, removal through hydroxyl radicals accounts for 90% of the sink strength, with the other 10% made up by soil uptake and other photochemical losses in the atmosphere. The removal of methane through hydroxyl radicals involves a rather complex chain of chemical reactions in which the presence of nitrogen oxides (see Chapter 11) is also important (Lelieveld et al., 1998).

But let us now discuss one of the major pathways of methane production: the breakdown of organic material. Since aerobic and anaerobic decomposition share a number of important features, we will discuss them here in a similar framework.

10.3 Decomposition

In the past, soil carbon modellers relied on a separation of labile and recalcitrant carbon into different pools, each with a specific turnover time and, importantly, a specific temperature sensitivity. 'Recalcitrant' and 'labile' were defined as intrinsic qualities of a molecule. This picture is slowly changing. The newly developed conceptual models rely more on the interaction of the microbial community with the substrate (Lehmann & Kleber, 2015) and soil structures.

10.3.1 Aerobic decomposition

Fundamentally, biomass or organic matter consists of a large variety of polymers. When biomass dies, soil fauna, such as worms, play a role in moving the organic material deeper into the soil. Once in the matrix, the polymers are broken down by exo-enzymes, enzymes operating outside the cell walls of the microbes. These exo-enzymes break down molecules of large molecular weight, such as carbohydrate lipids and proteins, into smaller components (Figure 10.2). The components of lower molecular weight become progressively more soluble in the soil matrix as they become oxidized. The soil matrix is best defined here as the total aggregate of soil particles, living organisms and organic material in various stages of decomposition. 'Soluble' in the soil matrix then applies particularly to absorption and desorption processes on and in this aggregate. Further interaction with the soil aggregates and mineral surfaces can protect the carbon from further decomposition and breakdown. This can happen when larger molecules are broken down into smaller ones, such as monomers like amino acids and monosaccharides, but the process also applies to larger molecules.

Importantly, whether a molecule becomes recalcitrant or not thus depends not only on its molecular structure but also on its position in the soil matrix: fixed or soluble. In essence, this is a scheme where polymers are progressively broken down into smaller components, whereby these components may go further down the chain, or get 'locked up' in the soil. The monomers can then be further oxidized to carbon dioxide. This conceptual model does not allow for the creation of specific recalcitrant pools, an essential

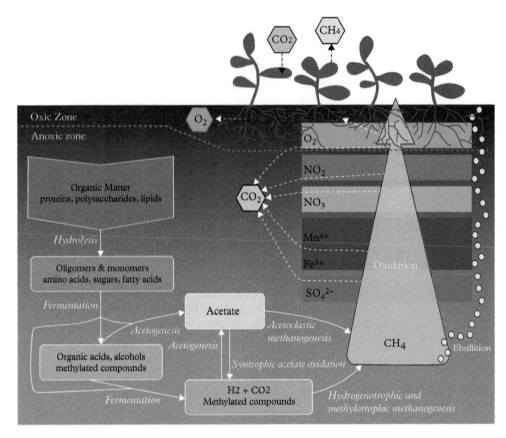

Figure 10.2 *Metabolic pathways for the degradation of organic matter from large polymers to carbon dioxide and methane. Microbial conversion processes are shown in italics. The methane diffuses upwards through the soil/sediment layer, where it can be oxidized by (an)aerobic methanotrophs. A distribution of available electron acceptors based on their electron potential is also shown on the right; this will vary between different environments. (After Dean et al., 2018)*

difference with the old standard model on which the soil component of most of our Earth system models are still based.

10.3.2 Anaerobic decomposition

The final step in the anaerobic degradation of organic matter is methane production. The labile compounds are broken down as in aerobic decomposition into monomers in the first step of decomposition: amino acids, fatty acids and monosaccharides (Megonigal et al., 2014). These monomers can then be converted into carbon dioxide or a combination of carbon dioxide and methane. The conversions to carbon dioxide occurs happily under aerobic conditions but, under anaerobic conditions, fermentation takes over and a consortium of bacteria is first required to degrade the monomers. This process usually includes one or two steps of fermentation to produce molecules with low weight such as volatile fatty acids or alcohols. When these are not directly converted after the first step, or convertible to methane or carbon dioxide, a second fermentation breaks them down into even smaller molecules, after which they can be converted into methane and carbon dioxide, or carbon dioxide only (Megonigal et al., 2014). In Figure 10.2, we see the different endpoints of anaerobic decomposition: either methane and carbon dioxide through the acetate pathway (eqn (10.1a)), or methane alone when carbon dioxide binds with hydrogen (eqn (10.1b)). As mentioned, not all methane produced in this way reaches the atmosphere, because it is easily oxidized into carbon dioxide during its transport to the atmosphere. This is a very important issue, both on land and in the ocean that we will come back to a number of times in this chapter.

Methanogenesis is regulated by the oxygen concentration, pH, salinity and temperature. Next to these, nutrients play a role and, of course, the substrate has to be available in the form of degradable organic matter. Because oxygen has such a strong control on methanogenesis, the absence of atmospheric air is a prerequisite. A very useful proxy for the presence of oxygen is the depth of the water table and, hence, strong relations have been found between methanogenesis and the water table. Micronutrients such as nickel, cobalt, sodium and iron also appear to be important to the functioning of methanogens. Absence of these micronutrients will hinder the production of methane.

Methanotrophic bacteria convert methane to carbon dioxide in the oxic zone by using oxygen (Figure 10.2). This was long thought to be the only thermodynamically attractive reaction. However, sulphate, iron and nitrogen oxides (Figure 10.2) can also act as electron acceptors. In the ocean, sulphate reduction is probably responsible for about 90% of the oxidation in the methane produced there. Sulphate reduction is performed by bacteria and anaerobic methanotrophic archaea (Dean et al., 2018). Methanotrophic bacteria associated with the roots of *Sphagnum* mosses have been found to decrease the amount of methane reaching the atmosphere by 98% (e.g. Raghoebarsing et al., 2005). These bacteria effectively use the methane as input for their carbon requirement, thus recycling the methane within the system and preventing it from reaching the atmosphere.

A common observation is that methane production declines with depth. This can be explained by the abundance of labile substrates near the surface and increasingly recalcitrant carbon deeper down in the profile—noting the remarks made earlier about the meaning of 'recalcitrant' in the context of the soil matrix. Near the top, freshly produced organic material, root exudates and newly produced plant detritus are generally more rapidly converted through the acetate pathway given by eqn (10.1a) while, deeper down, methanogenesis becomes more dependent on eqn (10.1b).

10.3.3 Rates of decomposition

Arrhenius developed the following equation for the relative rate of decomposition, V_{max}:

$$V_{max} = \alpha e^{-\frac{E_a}{RT}} \tag{10.2}$$

where α is a fitting parameter, R is the universal gas constant, T is the (absolute) temperature and E_a is the activation energy. The relative rate V_{max} provides the maximum rate given a certain temperature and sufficient availability of resource material. The activation energy signifies the 'push' that chemical reactions often need to get going; the exponent signifies the fraction of molecules that are available for a given temperature and have enough energy to cross the activation energy barrier and react. Importantly, reactions with high E_a, say, decomposition of more recalcitrant carbon, will have higher temperature sensitivities than reactions that have lower E_a, and so respire more labile carbon (Davidson et al., 2006). However, eqn (10.2) tells us nothing about the real or absolute respiration rates in the environment.

We thus need to find a way to incorporate resource availability. This can be achieved by including enzyme kinetics from the Michaelis–Menten equations. These equations originated in biochemistry and describe the breakdown velocity R in terms of substrate availability:

$$R = V_{max} \frac{[S]}{K_M + [S]} \tag{10.3}$$

In this equation, K_M is the Michaelis–Menten constant, that is, the substrate concentration at which $R = V_{max}/2$, and $[S]$ is the substrate concentration.

Temperature is a critical environmental constraint on both respiration fluxes and methane production. This sensitivity is often expressed as Q_{10}, which is the increase in the rate of production of a substance following a 10 °C increase in temperature. A $Q_{10} = 2$ thus signifies a doubling of the rate. The Q_{10} parameter plays a fundamental role in modelling decomposition. The values of Q_{10} in methane production and aerobic decomposition obtained from various experiments vary, from slightly less than 2 to about 4 or 5. However, the often-used value of 2 has neither theoretical nor experimental backing.

We can explain part of the Q_{10} variability by using the Arhenius equation (eqn (10.2)) and the Michaelis–Menten equation (eqn (10.3)). When the substrate S is freely available, its value in eqn (10.3) becomes much larger than K_M, so the rate is dependent only on V_{max}; V_{max} typically increases with temperature up to an optimum temperature beyond which the enzyme starts to deteriorate and, consequently, V_{max} declines rapidly. Therefore, when S is abundant and the temperature does not exceed the optimum, the reaction rate approaches V_{max}, as quantified by eqn (10.2). However, when S is low, K_M becomes important. Because the K_M of most enzymes also increases with temperature, the temperature sensitivities of K_M and V_{max} can neutralize each other (Davidson et al., 2006). This can create very low apparent Q_{10} values. These opposing sensitivities may be responsible for the large range of Q_{10} values found in the literature.

Soil moisture is another critical factor in decomposition kinetics under aerobic conditions. Soil moisture is important in any model describing respiration as, in the absence of soil moisture, decomposition slows down dramatically.

10.4 Methane sources

While methane emissions have a very large anthropogenic component (Figure 10.1), we concentrate here on the contributions of the natural environment, because these are likely to have played a role in some of the geological climate switches. It is, however, important to realize that a large component of the current total methane emission has anthropogenic origins, which makes methane a key gas for emission mitigation, given its global warming potential of 28.

10.4.1 Methane in wetlands

Wetlands are the main natural contributor to the global total of methane emission (see Figure 10.1). Once carbon is acquired through photosynthesis and sequestered into biomass and soil, permanent water logging prevents oxidation and slows down decomposition; but, importantly, it also facilitates the production of methane. Tropical wetlands are currently the biggest source (50%–60%), followed by northern wetlands (35%) and emissions from temperate wetlands (5%; e.g. Bloom et al., 2010).

Methane emissions from wetlands are difficult to quantify and also difficult to predict for the future. Part of this difficulty lies in the fact that the fluxes can use three different pathways: diffusion through the matrix, ebullition and transport through the vessels of vascular plants. This often leads to the association of high emissions with vascular plants (e.g. *Carex*). Transport through plants is a very efficient way to avoid oxidation during transport.

Forested wetlands (swamps) and freshwater marshes dominate the tropical wetlands. The conditions for methane production in these environments are very good: high productivity is achieved through high temperatures and abundant rainfall. It has been shown that trees can conduit up to 60%–90% of the ecosystem methane flux through their stems (Pangala et al., 2012). It is further thought that the variable extent of tropical wetlands, linked in part to the seasonal dynamics of river floodplains, plays a significant role in the year-to-year variability of atmospheric methane growth rates. Figure 10.3 summarizes the transport possibilities for methane in wetlands.

Temperate and northern wetlands between latitudes 40°–70° N are thought to contribute an amount roughly equal to half of the tropical wetland emissions. Their cold and wet environment is the main factor for this lower rate. Peatlands are the major source of temperate wetland emission. They comprise ombrotrophic peatlands (bogs) that receive water and nutrients from rain, and minerotrophic peatlands (fens) that receive their water and nutrients from underlying groundwater. The latter tend to be more nutrient rich. Optimal water tables for methane production are generally below the peat surface in bogs, above the surface in nutrient-rich fens and close to the surface in nutrient-poor peatlands. When these peatlands are drained, as many have been in the past to provide fuel, oxygen provides the source for aerobic decomposition and generally turns drained peatlands into sources of carbon dioxide with much reduced methane emission. This burning-off of peatlands can lower the surface by several metres over several tens of years. In low-lying areas such as in the Netherlands and in lagoons, this exacerbates the impacts of global sea level rise.

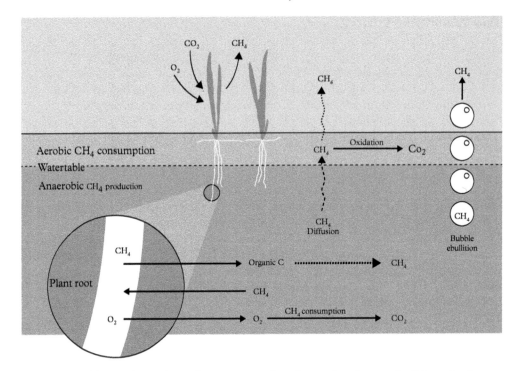

Figure 10.3 *Three pathways for methane transport through vascular plants and soil in wetlands: diffusion through soil, transport through plant vessels, and ebullition. (After Arctic Monitoring and Assessment Programme, 2015)*

10.4.2 Aquatic methane

Lakes and rivers are globally important sources of methane, with a high-end estimate of the flux coming to 122 Tg C yr^{-1}. The uncertainty in this estimate, however, is very large, with estimates of the flux ranging from 60 Tg C yr^{-1} to 180 Tg C yr^{-1} (Saunois et al., 2016). In Figure 10.1, they are shown as part of 'Other natural sources'. Very small ponds appear to contribute disproportionally to this flux (Holgerson and Raymond, 2016), and adequate characterization of these and the estimation of the effects of ebullition hinders the reduction of this uncertainty. In fresh water, methanogenesis through acetate formation (eqn (10. 1a)) is the dominant pathway while, in marine environments, hydrogenotrophic (carbon dioxide and hydrogen; see eqn (10.1b)) is the dominant pathway. Little of the methane produced in aquatic sediments, however, reaches the atmosphere. In marine sediments, anaerobic oxidation with sulphate consumes more than 90% of the methane produced. In freshwater environments, methane consumption takes place at the water–atmosphere interface. In general, less saline environments are more conducive to methane escape than strongly saline environments. The presence of sulphate dominates this distinction.

10.4.3 Methane in permafrost

Permafrost soils contain an estimated 50% of the total global soil carbon stock (1330–1580 Pg C). Its occurrence is shown in Figure 10.4 (see also Chapter 8 and Hugelius et al., 2014). Permafrost is defined as soil, sediment or rock that is permanently exposed to sub-zero temperatures for at least two consecutive years. Figure 10.4 shows the distribution of the carbon in the first 3 m. This immense store of carbon is vulnerable to degradation when conditions change, such as when temperature rises. Exposure of previously frozen material to either oxygen-rich or anaerobic conditions makes this pool vulnerable to microbial degradation. Methanogenic bacteria thrive in permafrost conditions; their activity has been measured even at temperatures of −20 °C! Acetate and hydrogen are considered the most significant substrates for methane production. Thaw and subsequent dynamics of the active layer of permafrost can increase the relative diversity, abundance and activity of methanogens. Aerobic methanogens can consume up to 90% of the methane produced in permafrost soils. Increasing thaw generally also initiates a shift in the diversity of the methanogenic microbial populations (Dean et al., 2018).

Changes in hydrology are, however, just as important as changes in temperature, because changing snow patterns provide melting ice and thus the anoxic environment required for methanogenesis. Thawed organic material can be deposited, either at small scales in lakes and ponds or at larger scales in the continental coastal zone. In both cases, the material may be subjected to microbial decomposition. Thus, unfrozen lake sediments in thermokarst lakes (taliks) may generate methane. It is debatable whether lake area is expected to increase when climate changes and, importantly, what effect this may have on future methane emissions. Van Huissteden et al. (2011) showed that methane emissions from lakes may increase through temperature and precipitation increase. However, drainage of thermokarst lakes does then provide an upper limit to the expansion of lake area and subsequent methane emission from Arctic lakes.

10.4.4 Methane in hydrates

Under certain high-pressure and low-temperature conditions, methane can be converted into a solid, ice-like substance called methane hydrate or clathrate. Methane captured in the hydrate crystal lattice is usually of biogenic origin and from layers deeper in the sediments. Methane hydrates can occur in areas where cold bottom water is present, with the overlying sediments and the water column providing the required pressure. Globally, 99% of all methane hydrates occur in continental margins. In permafrost areas, the lower temperatures may allow them to occur at shallower depths—shallower than the common 300–500 m at which they are found in non-permafrost areas. Figure 10.5 shows their distribution and transport pathways in the Arctic Ocean. The depth at which methane hydrates can form is known as the gas hydrate stability zone. Recent estimates of the amount of carbon involved are in the range of 1800–2400 Pg C (Kessler, 2017).

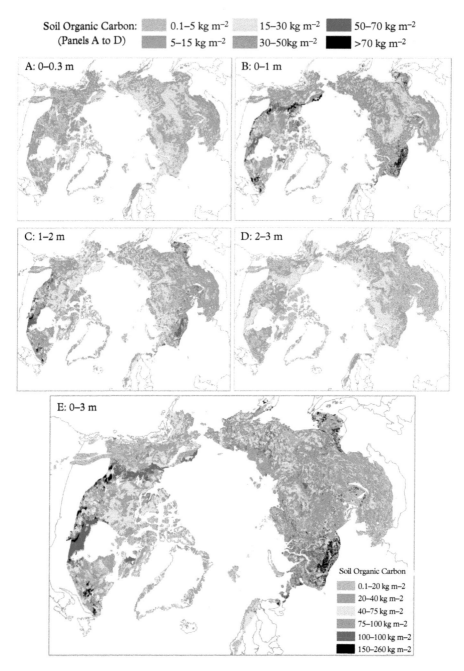

Figure 10.4 *Estimated carbon content (in kilograms per square metre) in permafrost soils. (After Hugelius et al., 2014)*

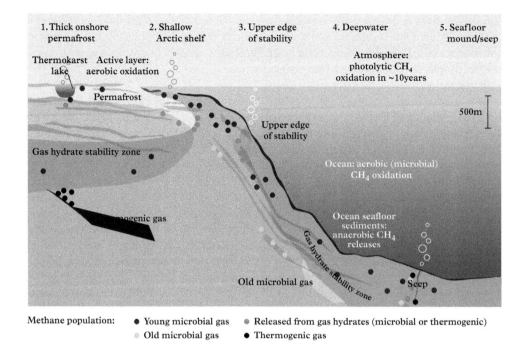

Figure 10.5 *Methane hydrate distribution in the Arctic Ocean. (After Arctic Monitoring and Assessment Programme, 2015)*

Destabilization of the gas hydrates is thought to be increasing, particularly in shallow regions in the Arctic Ocean where temperature is increasing. The East Siberian Arctic Shelf looks particularly vulnerable, although the precise magnitude and upscaling of individual measurements remains a problem. The estimates of emissions vary from 1 Tg C yr^{-1} to 17 Tg C yr^{-1}, a small amount compared to wetlands or even lake and river emissions. This is partly because the methane is released at depths well below the surface and, consequently, a long pathway exists where the methane is prone to oxidation to carbon dioxide. In general, this is one of the reasons why the total methane production of the global ocean is a very small contribution to the overall budget. That being said, one of the enigmas of ocean methane is the so-called ocean paradox, whereby supersaturation of methane occurs near the surface. The reasons for this so far have appeared illusive.

10.5 Methane in the geological perspective: the faint sun paradox

Methane together with carbon dioxide probably played a big role in providing the greenhouse warming required to offset the low radiance of the 'faint sun' (e.g. Feulner, 2012).

The earlier Sun is estimated to have radiated at 70% of its current value during the Archean. In principle, without any greenhouse gas effect, this would imply a surface temperature of 233 K. Remember the large difference of 33 °C between a greenhouse-gas-free world and the current one, as explored in Chapter 4? Without methane, the surface temperature of the early Earth could have been 22 °C lower than with methane.

Additional greenhouse gases like methane would have kept the oceans liquid at a time when the radiance of the Sun, most likely, was not enough to do so. The development of this scenario is given in Figure 10.6, where the evolution of carbon dioxide, methane and oxygen is shown (Nisbet & Nisbet, 2008). It is worth noting that the original idea for this figure came from James Lovelock, the originator of the Gaia theory that describes Earth, including the atmosphere, as a single big organism, where life plays an active role in the atmospheric composition and the maintenance of the system. Around 3.8 Gyr ago, the first archaean methanogens (methane producers) appeared. These archaea were probably in possession of an early variant of the rubisco enzyme, rubisco III. Rubisco, in its modern form (rubisco I) is the key enzyme in the Calvin cycle of photosynthesis, taking carbon dioxide out of its environment and converting it to sugars and oxygen in plants (eqn (1.1)). Rubisco III, however, is an enzyme that requires anoxic conditions. With the appearance of the first methanogens, the methane concentration could have risen quickly

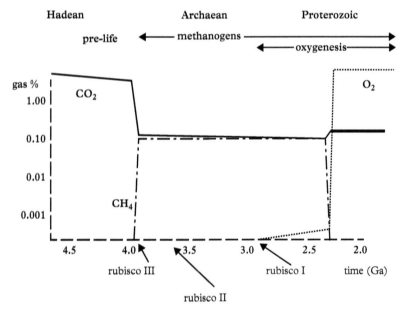

Figure 10.6 *Interpretation of the geological records in terms of methane, carbon dioxide and oxygen. Note the appearance of oxygen after the Great Oxidation Event at 2.3 Gyr, and the different times at which variants of the rubisco enzyme appeared. Rubisco III is required for anoxic methane production. (After Nisbet and Nisbet, 2008)*

around 3.8 Gyr ago. When rubisco I, the oxic variant appeared, around 2.9 Gyr ago, the story of methane dominating the greenhouse gas composition of the atmosphere stalled, as oxygen converts methane very quickly into carbon dioxide with the help of hydroxyl radicals. With abundant oxygen in the atmosphere, methane was again relegated to the position of a trace gas. Carbon dioxide, methane and oxygen are thus intricately linked in Earth's geological history. Together with nitrous oxide (Chapter 11), methane and carbon dioxide are sometimes called 'Earth's managing greenhouse gases' (Nisbet & Nisbet, 2008).

10.6 Methane in glacial cycles

Unfortunately, no reliable proxies exist that allow us to track the methane concentration down into geological time, other than the methane trapped in ice cores. Methane concentration in ice cores has ranged from approximately 320 ppb up to 780 ppb. Warm interglacials have high values, and cold glacials have low values—very much like carbon dioxide (Quiquet et al., 2015). The explanation of these differences is not without its problems. As in the case of carbon dioxide (see Chapter 9), the atmospheric concentration of methane is a balance between sources and sinks, both of which show a sensitivity to changes in climate. In the case of methane, this can be a change in the sink capacity, a change in the source capacity or a change in both. The sink capacity is determined by the availability of the hydroxyl radical that forms after ozone is split and has reacted with water. This radical then convert methane into carbon dioxide and water.

The source capacity has to be the largest natural source: wetlands. Early estimates suggested a 44% decrease in wetland emissions during the Last Glacial Maximum as a consequence of the much colder and drier climate and extensive ice sheets. Such a large change would be able to explain the observed rise of the Last Glacial Maximum to pre-industrial methane concentrations of 350–700 ppb in ice cores. However, when bottom-up models were used, these were generally not able to explain the increase in emission—an increase in sink strength was also needed. A recent review (Quiquet et al., 2015) concluded that it is possible to achieve the required 49% reduction in methane level between the pre-industrial period and Last Glacial Maximum with a reduction of wetland emission by 43%. Realistic estimates vary from 38% to 43% and, with a concurrent reduction of between 30% and 35% in other sources, it may also be possible to achieve the reduction without major changes in the sink and the lifetime of methane.

Such an explanation makes the proposed additional contribution from clathrates obsolete. This so-called clathrate gun hypothesis assumed that warming during warm interstadials (Dansgaard–Oeschger events) would destabilize the clathrates at the continental shelves. There is, however, no isotopic evidence for this release; in contrast, isotopic analyses of hydrogen in methane suggest that the increased methane did not originate from destabilized methane hydrates (Kessler, 2017). There is thus no need to invoke the clathrate gun hypothesis or sudden geological emissions to explain some of the sharp temperature and methane increases in the last interstadials. Increased methane emission by a wetter environment and higher temperatures is likely to be a sufficient explanation, although there remains a large uncertainty, particular about the size of the wetland areas.

10.7 The anthropogenic perturbation

Figure 10.7 shows the evolution of the atmospheric methane concentration since the late 1980s. First, it is noteworthy that the current global mean concentration is about 1850 ppb, the maximum obtained from samples of the glacial–interglacial atmosphere trapped in ice cores being 780 ppb. This substantial increase is unequivocally due to humans, particularly agriculture and fossil fuel burning (see Figure 10.1). It should also be noticed that the atmospheric growth rate (the slope of the curve) declined from about 1990, with an even larger decline up to 2006, and since then increased. The causes of this variability are still unknown and subject to much debate. This generally highlights the uncertainty involved in putting together the global methane budget. It is important to emphasize both the relevance and the causes of the controversy. Methane is a much stronger greenhouse gas than carbon dioxide, and any lack in understanding of the causes of its variability hinders understanding of mitigation options (see Chapter 13).

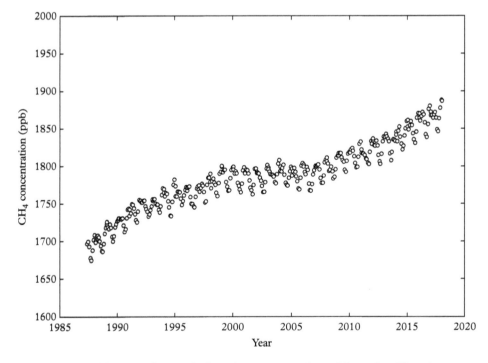

Figure 10.7 *Monthly averaged atmospheric methane concentration at Mauna Loa. These data are provided by NOAA. Principal investigators include Ed Dlugokencky, NOAA. (Based on Dlugokencky et al., 1995)*

One of the important factors in explaining the rise is the change in $\delta^{13}C$ of methane (Nisbet et al., 2016). A decrease in global values of $\delta^{13}C_{methane}$ of about $-0.24‰$ has been observed from measurements of the global atmospheric composition network (see Chapter 3). This points to a biogenic origin of the increased methane, rather than a fossil fuel source, as gas from the former would be more depleted of $^{13}C_{methane}$. The increase in emissions and associated decline in $\delta^{13}C_{methane}$ is, in principle, consistent with increases in anthropogenic methane (in particular, through agriculture), increases in Arctic emissions or increases in wetland emissions (in particular, tropical wetland emissions). The latter are generally thought to comprise the main contributor to the interannual variability, but unfortunately, few measurements exist around the tropics, and satellite data are still not of sufficient reliability and coverage to help out here. Similarly, and again emphasizing the uncertainty in these bottom-up estimates, for the Arctic, using several sources of wetland extent with a bottom-up methane model, Petrescu et al. (2010) obtained very large differences between estimates when using different wetland extent maps. The variation in annual wetland extent is thus a key factor in determining the Arctic emissions, and this also holds for the tropics, which appears to drive much of the interannual variability.

So, what is causing the increase after 2007? The isotopic signal rules out a significant fossil fuel contribution, such as, for instance, that obtained from leaking gas in Siberia or from fracking in the United States: neither of the two is fully consistent with the decline in observed $\delta^{13}C_{methane}$. It appears that a recent decrease in global biomass burning, although relatively small in the overall methane budget but with a disproportionally large impact on the $\delta^{13}C_{methane}$ budget, may be able to bring the various explanations together, while remaining plausible within the isotope budget (Worden et al., 2017). This argument does not require us to invoke additional rises in the contributions from fossil fuel, or a decrease in the hydroxyl-radical concentrations enhancing the sink of atmospheric methane. Overall, it would thus appear that natural variability in tropical methane emissions is the main cause, with the precise source being somewhat more uncertain. However, we need to be careful to rule out completely changes in the concentration of hydroxyl radicals that drive the oxidation capacity of the atmosphere. Elucidating changes in any biogeochemical reservoir will always require us to carefully evaluate both the source (input) and the sink (output).

10.8 Future methane emissions

The variability raises an important issue in the context of past and future climate change (e.g. see Nisbet et al., 2016). If it is true that the increase in methane is related to changes in precipitation, which drive the extent, and temperature, which increases the rate, it raises the question of how these changes relate to those in the past as found in ice cores. In those changes in glacial–interglacial cycles, increased methane concentrates usually relate to fast climate changes. No matter what the response of the current climate will be, the sobering thought is that the observed increase in the last decade is substantially larger than the range observed during the previous two millennia.

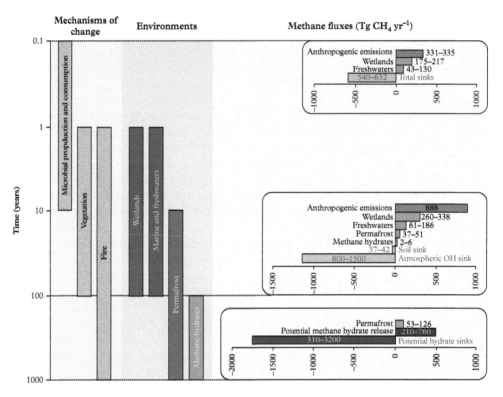

Figure 10.8 *Estimated timescales of potential methane climate feedbacks; the timescale indicates the response time of the mechanisms of change to climate forcing, and the immediacy of the impacts of these changes to methane emissions from different environments. (After Dean et al., 2018)*

Thus, as important as understanding what causes the current increase is the understanding and prediction of what we may expect when our climate continues to warm. Given the global warming potential of methane, any increase in emissions has the potential to further enhance the warming, so a good knowledge of emissions sources and sinks and their sensitivity to climate, primarily temperature change, is important. Figure 10.8 shows the estimated contribution of several key methane emissions sources under future climate change. The estimates are obtained from a variety of sources, mostly obtained with a high-end climate change scenario, such as RCP 8.5 (Dean et al., 2018).

The graph shows the timescale of the crucial sources, showing that we first expect microbes to adapt and change, and later vegetation and associated fires. Around the same timescale, the extent of wetlands and the marine and freshwater sources may change (1–100 years). Processes such as permafrost degradation and methane from hydrates operate at much longer timescales, with the hydrates expecting to have their primary sensitivity at a 1000-year timescale.

In the horizontal insets, the expected range of fluxes is given. At the top right, the current fluxes are given. Wetlands remain an important source for increase under climate change but, importantly, the sink capacity is also likely to change. How this will eventually affect the net balance is still an open question. At longer timescales, hydrates become important, but they are unlikely to contribute much in the next 100–200 years. The graph also show that anthropogenic emissions remain a substantial part of the projected budget. Finding reductions in these methane emissions may turn out to be the key to successful mitigation of climate change.

11

The Nitrogen Cycle and Climate

11.1 The nitrogen cycle

After carbon, nitrogen is the most important nutrient on Planet Earth. Similar to the carbon cycle, the emergence of life on the planet drastically changed the nitrogen cycle; similar to carbon, human intervention in the modern nitrogen cycle has put its operation outside past boundaries. The impact on climate is both direct, for example through the greenhouse gas nitrous oxide, and indirect, through its interaction with the carbon cycle and the formation of ozone and aerosols. Nitrogen is needed to build amino acids, proteins and enzymes such as rubisco (see Chapter 10), which are crucial for photosynthesis. In contrast to the carbon cycle where the dominant fluxes such as photosynthesis are controlled by vegetation, the nitrogen cycle is dominated by microbial activities and recycling. This makes it a much more complex cycle to investigate. Let us start by illustrating this complexity.

Novel gene-reproducing and analysis techniques make it possible to study the complexity of microbial activity in the subsurface and sediments (Anantharaman et al., 2016). A recent analysis found a staggering 1297 clusters of genomes representing distinct microorganisms in the soil. All of these contribute some chemical step in the cycles of nutrients and a large percentage of genomes found were identified as belonging to a new phylum. It is a complexity based on chemical handoffs that are common (a species performing one chemical reaction, on whose product the next species can then act) and multiple metabolic abilities of organisms that perform maybe only one type of redox reaction but can function in different biogeochemical cycles. This 'plug and play' strategy enables an enormous variety of microbial system configurations and probably enhances the ecosystem's resilience in the face of perturbations (Anantharaman et al., 2016). The question how this vastly complex, still largely unknown system responds to environmental change is key to our ability to predict the future.

Figure 11.1 presents a simplified view of the nitrogen cycle, showing the most important conversions and forms of nitrogen. Most of the nitrogen is present in gaseous form as dinitrogen in the atmosphere. Prokaryotic organisms from Archaea and Bacteria are able to form ammonium from the inert gas. Nitrogen-fixation organisms appear very broadly across a range of environments, from soils, oceans and lakes to sediments and termite guts. Using the enzyme nitrogenase, they are able to break the strong triple bond

Biogeochemical Cycles and Climate. Han Dolman, Oxford University Press (2019). © Han Dolman.
DOI: 10.1093/oso/9780198779308.001.0001

between the two nitrogen atoms forming a molecule of dinitrogen. Specific species do this breakdown in both aerobic and anaerobic conditions. They can appear either as free-living organisms or in specific symbiotic relationships with plants. In order to break the very strong triple bound in dinitrogen, energy is required. The energy is provided by the plant roots in the form of sugars. The plants benefit from the nutrients they get in return. This is a clear interaction between the carbon cycle and the nitrogen cycle: the sugars are produced through photosynthesis, for which nutrients are needed. Nitrogen fixation is an extremely important and essential process, because plants and other organisms generally cannot use dinitrogen, but rely on ammonium or its oxidized forms, nitrite and nitrate.

The microbes executing the next series of conversions in the cycle are also part of the Bacteria and Archaea domains, although our knowledge of the latter is limited. This part of the nitrogen cycle is called nitrification. Aerobic ammonia oxidizers convert ammonia to nitrite and they are autotrophs: they also use ammonium as an energy source, rather than light, to convert carbon dioxide into organic carbon. The next step in the cycle, nitrite oxidation, is executed only by aerobic bacteria.

From nitrite on, there are two fundamentally different chemical pathways to produce the end product, dinitrogen. The classic way is the conversion of two molecules of nitrite to dinitrogen and is called denitrification (Figure 11.1). In the denitrification pathway, nitrite is converted to dinitrogen by anaerobic chemoautotrophic bacteria. During this process, nitrous oxide is produced. Nitrous oxide is an important greenhouse gas: its

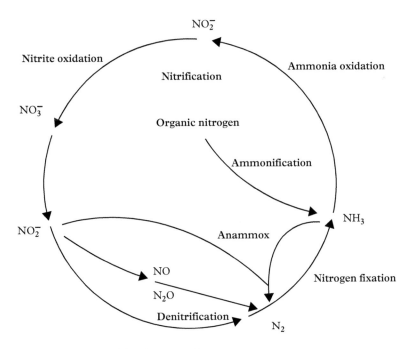

Figure 11.1 *Simplified view of the nitrogen cycle.*

global warming potential is 268 and 298 for a 20-year period and a 100-year period, respectively (Ciais et al., 2013). The lifetime of nitrous oxide in the atmosphere is relatively long: 121 years ($\pm$10). For climate, nitrous oxide is the most important element of the nitrogen cycle. As for carbon dioxide and methane, its levels have steadily increased in recent years (see Figure 11.2). But the other forms of reactive nitrogen, collectively referred to as N_r and defined as all forms of nitrogen except dinitrogen, are also important because they easily link with the carbon or phosphorus cycle (see Chapter 12) or aerosols (see Chapter 5), and impact climate through those paths.

The other dinitrogen-producing pathway was discovered more recently in 1999. This new pathway, by which the nitrogen atoms of nitrite and ammonium are combined to form dinitrogen was first shown to operate in the anaerobic environment of a wastewater bioreactor in the Netherlands. The microbe responsible was subsequently detected in sewage treatment plants all over Europe. It is interesting to note that the possible existence of this pathway was first proposed in the 1930s, based on thermodynamic and chemical reasoning, but the actual microbe that performed the reaction could not, at that time, be isolated. The process converting nitrite and ammonium into dinitrogen is called Anammox, for *anaerobic ammonium oxidation*. It is now thought that the Anammox pathway plays a significant role in the nitrogen cycle, particularly in the marine nitrogen

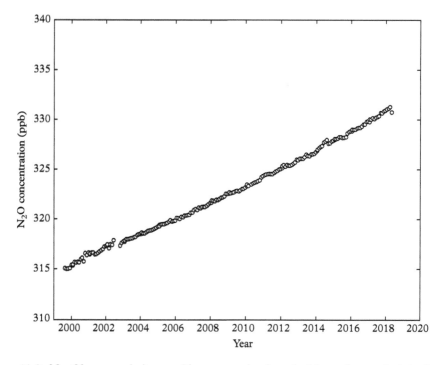

Figure 11.2 *Monthly averaged nitrous oxide concentration from the Mauna Loa station's in situ measurements. Nitrous oxide data from the NOAA/ESRL halocarbons in situ programme.*

cycle. In 2003 its existence was confirmed outside the bioreactor environment, in the world's largest anoxic sea basin, the Black Sea. Anammox is thought to be responsible for about 20%–30% of the marine conversion of nitrite to dinitrogen (Kuypers et al., 2003).

11.2 Human intervention: the Haber–Bosch process

So far, we have discussed the natural, pre-1900 nitrogen cycle. The nitrogen cycle changed dramatically by humans in their pursuit of producing food and particularly when Fritz Haber discovered the process by which dinitrogen, under high temperature and pressure, together with hydrogen could be converted to ammonia. He patented this process in 1908. Carl Bosch subsequently scaled up the process for use in industrial applications. Since then, the process is generally known as the Haber–Bosch process. Their discovery was extremely useful: first, it allowed the large-scale production of artificial fertilizer, setting off the population explosion in the twentieth century, especially after the Second World War. It also produced the basic building material for explosives (precursors of TNT). Paraphrasing Erisman et al. (2008), the importance of

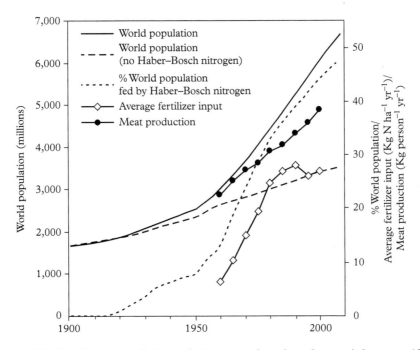

Figure 11.3 *Trends in human population and nitrogen use throughout the twentieth century. (After Erisman et al., 2008)*

Haber's discovery cannot be overestimated—'as a result, billions of people have been fed but, at the same time, millions of people have died in armed conflicts over the past 100 years'.

Figure 11.3 shows the growth of the human population, together with an estimate of the contribution of the Haber–Bosch process to providing food for humans. While, of course, these are estimates, it is telling to the importance of the process that, without the use of artificial fertilizer, the human population would have been roughly 50% of what it is today. That being said, only small fractions of nitrogen fertilizer end up in plants; for instance, in 2005, from the approximately 100 Tg N used in global agriculture, only 17 Tg N was consumed by humans in crop, dairy and meat products. The remaining part either was denitrified back to dinitrogen or ended up as N_r (reactive nitrogen) in the environment, where it creates havoc while cascading through the soil, water and ocean, from where it eventually may be returned as dinitrogen through denitrification or Anammox. One molecule of N_r can, in this way, have various adverse effects at multiple points in time and space, a phenomena termed 'the nitrogen cascade' (e.g. see Galloway et al., 2003). The environmental impacts of nitrogen are severe: loss of biodiversity, pollution of water and air and increases in the greenhouse gas nitrous oxide.

11.3 Atmospheric nitrogen

By far, the largest amount of nitrogen is found in the global atmosphere, of which 78% is dinitrogen. This is 99.99% of the nitrogen found in the atmosphere, with nitrous oxide contributing virtually all of the remaining 0.01%. The small remainder is made up by ammonia, ammonium and nitrogen oxides (either nitrogen monoxide or nitrogen dioxide). These are trace gases, of fundamental importance to atmospheric chemistry.

When compounds of nitrogen are involved, the atmospheric conversions become quickly complex. They can conveniently be divided into the impacts of reduced and oxidized forms. Figure 11.4 show the main conversions and original sources. Ammonia is emitted mainly from agricultural sources. In the atmosphere, it is transported and dispersed but, importantly, also scavenged as dry deposition (deposition of particles on surfaces under dry conditions, as opposed to wet deposition of precipitation containing dissolved forms). The part that remains in the atmosphere can form aerosol-bound ammonium ions in reactions with acid gases and aerosols (see Chapter 5). Aerosol-bound ammonium has a relatively long lifetime in the atmosphere and may be transported over long distances (>1000 km). Ammonium-containing aerosols are removed mostly by wet deposition onto vegetation or soil. This transport creates impacts of ammonium far away from the location of the original source.

Nitrogen oxides are mostly emitted to the atmosphere as either nitrogen monoxide or nitrogen dioxide and are commonly collectively referred to as 'NO_x'. The 'cleaning agent' of the atmosphere, the hydroxyl radical (see also Chapter 10), then reacts with nearly all of the nitrogen oxides to form nitric acid. Nitric acid has a short lifetime in the atmosphere because it is quickly taken up in aerosols, reacts with ammonia or is deposited under dry conditions on to the surface. Uptake in aerosols or formation of new aerosols

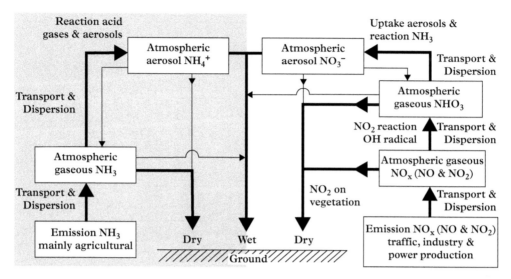

Figure 11.4 *Complexity of pathways of reactive nitrogen in the atmosphere. The left side of the figure illustrates the atmospheric pathways of gas phase ammonia (NH₃) and the aerosol phase ammonium (NH₄⁺). Dry is dry deposition; wet is wet deposition. The right side of the figure illustrates pathways for the nitrogen oxides. (After Hertel et al., 2011)*

by the reaction with ammonia leads to aerosol-bound nitrate. Also, nitrate-containing aerosols are scavenged by wet deposition. As with ammonium, the crucial relevance of this part of the atmospheric nitrogen cycle is the transport away from the emission sources and the conversions to aerosols. Eventually, the reactive nitrogen is deposited onto the land surface and vegetation, where it can then contribute as an important nutrient to plant growth or cause acidification of soils.

11.4 Terrestrial nitrogen cycle

Once deposited, the different forms of nitrogen become part of the terrestrial nitrogen cycle. Figure 11.5 shows a schematic of the terrestrial nitrogen cycle. It also shows the anthropogenic fluxes that have perturbed the natural nitrogen cycle since the invention of the Haber–Bosch process. The latter is particularly obvious in the arrows representing industrial and agricultural nitrogen fixation. For example, the deposition of nitrogen on land has doubled in size; similarly, the nitrogen outflow through rivers to the ocean has also almost doubled. Emission of NO_x and ammonia are also strongly affected. Although to a lesser extent, the nitrous oxide fluxes are also significantly affected.

Nitrogen cycling and storage depend very much on the type of land use and its history. (NB: The term 'land use' is reserved to indicate the use and management of land; 'land cover' is a physical classification of the land surface only). Agricultural systems

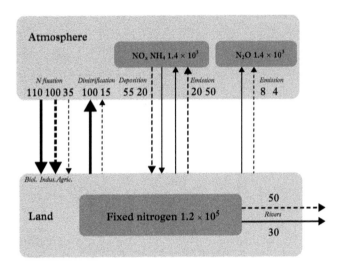

Figure 11.5 *Schematic of the terrestrial nitrogen cycle. Stocks are in teragrams of nitrogen; fluxes are in teragrams of nitrogen per year. Dashed lines represent fluxes induced by humans. Nitrogen transformations are shown in italics; stocks are shown in bold. (Values from Gruber & Galloway, 2008)*

where large amounts of nitrogen-containing fertilizer are being used, are generally (over)saturated in nitrogen, similarly with natural areas or forest that are in areas where large amounts of N_r is deposited. The impact of these additions may last long after the land use has changed. In contrast, northern forest may suffer from nitrogen limitation as the decomposition and the low temperatures limit recycling of nitrogen from organic material.

Figure 11.6 shows the fundamental differences between nitrogen cycling in natural vegetation, such as a forest, and that in an agricultural crop. The plant nitrogen stocks, in a way, are surprisingly similar between the two groups (note, however, that, in a temperate rather than a boreal forest as shown here, almost ten times more nitrogen is stored in the plants). However, there are very large differences in the other components, in particular, in the supply of fertilizer and the subsequent deposition of nitrogen. Quantitatively, all of the nitrogen fertilizer is exported from the agricultural crop through harvest, but deposition still adds to the stocks. Thus, the plant nitrogen uptake is an order of magnitude greater in the agricultural crop than in the forest.

This of, course, reflects the different life cycles of the two systems: nitrogen in forest soils is recycled during the lifetime of a forest while, in an agricultural crop, it is taken up in a single year. Importantly, and this is where the climate aspect plays a role, the large stock of nitrogen in the soil of an agricultural crop allows significant amounts of nitrates to be denitrified in the form of gases (see Figure 11.1), the production of which is minimal or absent in the boreal forest. In addition, in areas with high deposition rates and fertilizer use (e.g. the Netherlands), leaching of nitrogen from the soil into surface waters presents massive environmental problems. Thus, while the input into the boreal

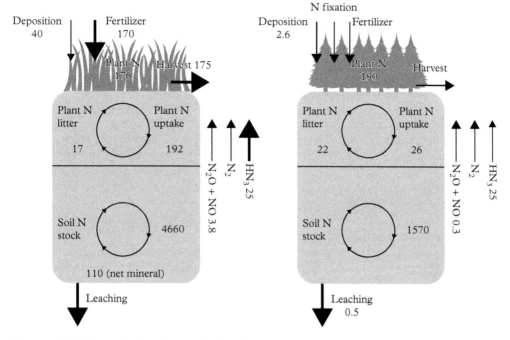

Figure 11.6 *Nitrogen budget for an agricultural crop in Belgium and a coniferous forest in Finland (in kilograms of nitrogen per hectare). Where no value is given, the stock or flux is not known. (Redrawn after Butterbach-Bahl & Gundersen, 2011)*

forest from biological nitrogen fixation and deposition almost matches the differences between nitrogen in litter and nitrogen uptake, in the agricultural crop there is an excess of nitrogen that is denitrified into gas, leaches into water or adds to the soil stock, with most of the soil nitrogen being fixed to organic material, where it either remains or is ultimately decomposed (see Chapter 10).

The input from biological nitrogen fixation, both globally and locally, has been difficult to determine experimentally but is estimated to be of the order of 1.5–2.0 kg N ha^{-1} yr^{-1} for boreal forest, 6.5–26.6 kg N ha^{-1} yr^{-1} for temperate forest and 2.3–3.1 kg N ha^{-1} yr^{-1} for natural grassland (although for managed grasslands, it is a factor of 100 more); for agricultural systems, it varies from about 30 N ha^{-1} yr^{-1} to 120 kg N ha^{-1} yr^{-1}, depending on the crop. It appears best to take the lower estimates as the more probable, based on the assumption that most of the experimental studies have been performed at locations where biological fixation was likely to play a role. However, the large range between various biomes certainly indicates the great variability in the real world and the difficulty of establishing a robust global estimate. Furthermore, we lack knowledge about the precise drivers of nitrogen fixation and the impact on nitrogen stores. In tropical forests that appear sometimes remarkably rich in nitrogen, with very speedy turnover rates, this has led to the so-called nitrogen paradox: how can we have this richness in the biome sustained only by nitrogen fixation that should become less when nitrogen is available (Hedin et al., 2009)?

11.5 Nitrous oxide

The dominant sources of nitrous oxide are microbial production processes in soils, sediments and water bodies. Agricultural emissions due to nitrogen fertilizer use and manure total 4.3–5.8 Tg N_2O–N yr^{-1} and emissions from natural soils are estimated at 6–7 Tg N_2O–N yr^{-1}. Together they represent 56%–70% of all global nitrous oxide sources (Butterbach-Bahl et al., 2013), the remainder being made up from fertilizer production and traffic and oceanic losses. The nitrous oxide emissions vary tremendously, both in space and in time, and most of our process knowledge comes from laboratory experiments rather than in situ observations.

Globally, the IPCC (Ciais et al., 2013) estimated in their Fifth Assessment Report that for the period 2006–11, soils under natural vegetation emitted 6.6–7.0 Tg N_2O–N yr^{-1} with the total natural sources, including production from lightning, amounting to 11 Tg N_2O–N yr^{-1}. Total anthropogenic sources amounted to 6.9 Tg N_2O–N yr^{-1} with agriculture contributing 4.1 Tg N_2O–N yr^{-1}. However, the uncertainties are very large: for instance, the range of the total anthropogenic estimate is 2.7–11.1 Tg N_2O–N yr^{-1}. In addition, 14.3 Tg N_2O–N yr^{-1} is removed by chemical reactions in the stratosphere. Emissions during 2017 are thought to be eight times higher than they were in 1990. The difference between emission and removal constitutes a positive growth rate in the atmosphere with associated increases in radiative forcing (see Figure 11.2).

Denitrification is a 'broad' microbiological process. The microbes responsible for it are from a wide range of phyla; most of them are from Bacteria but, more recently, archea have also been found to be involved. As with methane, soil moisture and temperature are the main controlling factors for emission of nitrous oxide from soils. Soil moisture determines the degree of oxygen (or lack there of) in the soil; temperature affects control through sensitivity of multiple decomposition processes down the denitrification chain. Generally, up to 80% moisture saturation of soil leads to nitrous oxide production. Above 80% saturation, the available substrate is converted mainly into dinitrogen. Seasonal and spatial dynamics of soil moisture, its occurrence already quite heterogeneous due to the erratic nature of rainfall, and temperature also impact nitrous oxide emission rates. For instance, temporary waterlogging, seasonal changes from drought to wetting as well as transient zones between upland and wetland soils can constitute so-called 'hot moments' and 'hot spots' for nitrous oxide emissions as they present ideal conditions for the transition from microbial oxygen to nitrate respiration (Butterbach-Bahl et al., 2013). Nitrous oxide reactions also exhibit hysteresis behaviour: nitrous oxide production during rising temperatures can follow a different trajectory when temperatures are falling, owing to faster depletion of substrates.

11.6 Nitrogen stimulation of plant growth

Nitrogen is an essential element in amino acids and plant enzymes. Since most soils are naturally limited by nitrogen availability in a plant-available form, increased availability, for instance, through enhanced deposition should in principle stimulate growth. The increase in the deposition due to human use can be calculated using a distribution of sources and an atmospheric transport model that takes into account the chemical transformations (e.g. those shown in Figure 11.4). Global nitrogen deposition is mapped in Figure 11.7 for both the pre-industrial period and 2000. The graphs show the deposition of reactive nitrogen and ammonium as well as dry and wet deposition. Five large areas can be distinguished that relate primarily to the anthropogenic, agricultural use of nitrogen and centres of industry and population: Europe, the United States, South East Asia, central Africa and south-east Latin America. Quite visible also is the increase in deposition from 1850 to 2000. Projections for the future suggest there will be an even larger deposition.

How much of this enhanced deposition has led to stimulation of forest growth, is highly controversial (Ehtesham & Bengtson, 2017). Estimates of Net Ecosystem Production being stimulated vary from at rates of 30–75 kg C per kilogram of nitrogen down to 16 kg C per kilogram of nitrogen. The first, high-end estimates are generally criticized because they lack ecological plausibility: the resulting carbon-to-nitrogen ratios are so much outside the common ranges that they may simply not be possible or sustainable. A meta-analysis showed the stimulation of above-ground woody production by values of around 4–5 for old forest and up to 20 for young forest. So, not only forest age but also productivity status matters, because low productivity forests respond more strongly than high productivity forests (Schulte-Uebbing & de Vries, 2017). The issue is further complicated by the fact that not only growth (Net Primary Production or Gross Primary Production; see Chapter 9) responds to nitrogen. It is also well known that decomposition is also, but sometimes negatively, affected by nitrogen fertilization (Janssens et al., 2010). This makes it hard to make robust estimates of the net result.

It is, however, possible to identify three potential mechanisms that could result in enhanced growth and carbon uptake by forests as a result of enhanced deposition. This is illustrated in Figure 11.8. The first is stimulation of photosynthesis (or Gross Primary Production) by the enhanced availability of nitrogen. There exists, in general, a clear relationship between foliar nitrogen and photosynthetic rates across biomes, as the photosynthetic machinery needs a lot of nitrogen. In young forests, the response of primary productivity to enhanced nitrogen is high; this is also the case in boreal forests, where nitrogen availability is the critical limitation for growth (Janssens & Luyssaert, 2009). In temperate forests, this enhancement is often less clearly visible, as nitrogen levels may have already reached saturation and further enhancement appears to have no effect on tree growth. Note that additional nitrogen loading can still have severe implications for the rest of the ecosystem and environment, including biodiversity, as leached reactive nitrogen further cascades through the system. The second mechanism is a shift in carbon allocation. Plants normally invest a substantial fraction of their photosynthetic

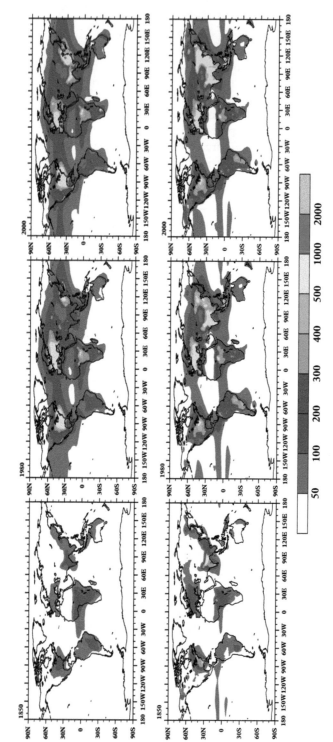

Figure 11.7 *Computed wet and dry total nitrogen deposition rates for nitrogen oxides (upper three panels) and nitrogen hydrides (lower three panels), for the year 1850, using an atmospheric transport model (left panels), and, for the year 2000, using the ensemble average of several models (right panels), in milligrams of nitrogen per square metre per year. (After Lamarque et al., 2013)*

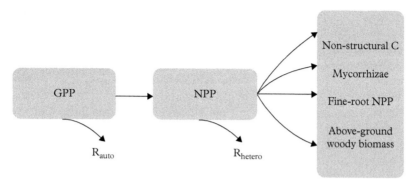

Figure 11.8 *Schematic of the carbon allocation and flows in forest: Gross Primary Production (GPP), Net Primary Production (NPP), autotrophic (plant) respiration (R_{auto}), Net Primary Production (NPP) and the allocation into non-structural carbon (C), mycorrhizae, fine-root NPP and above-ground woody biomass.*

production in roots and root symbionts such as mycorrhiza, which help in gathering nutrients. As nutrients become less limiting, the allocation of carbon can shift from fine roots (below-ground Net Primary Production) and mycorrhizal symbionts to the woody biomass without affecting overall primary production. In this case, this allocation mechanism in itself would lead to higher wood production per se, even if photosynthesis, for instance, were to show little response to the enhanced availability of nitrogen. The third mechanism is the little-understood process of respiration in the soil. In Chapter 10, we saw the complexity of the decomposition of soil organic matter. While it is known that enhanced nitrogen appears to slow down decomposition of organic material, how much of this remains as soil carbon is unclear. How all of these three processes feedback on the growth, if more nitrogen can be allocated to wood and if this ultimately represents a shift in nitrogen allocation within trees is largely unknown.

Based on a meta-analysis of the carbon-to-nitrogen ratios of trees, Schulte-Uebbing & de Vries (2017) estimate a global nitrogen-induced carbon sequestration in above-ground woody biomass production of 93–202 Tg C yr^{-1}. This is partitioned into the contributions from temperate forests (101 Tg C yr^{-1}) and boreal forests (35 Tg C yr^{-1}); tropical forests make up most of the remainder with a relatively small contribution (15 Tg C yr^{-1}). Adding an estimate of below-ground woody production yields a total stimulation of carbon growth of 177 Tg C yr^{-1} or 0.18 Pg C yr^{-1}. This is less than 10% of the current terrestrial sink (see Chapter 9).

11.7 Oceanic nitrogen cycle

The nitrogen cycle in the ocean is as complex as that in the terrestrial domain, with many forms of nitrogen, from reduced to oxidized, cascading through the system. The most ubiquitous component is, again, dinitrogen (see Figure 11.9). In the ocean, as on land, nitrogen-fixation organisms convert this into a form that is usable by other organisms (in the case of the ocean, by phytoplankton). In fact, the amount of biological nitrogen fixation in the ocean is equal in magnitude to that on land (see Figure 11.5). Interestingly, the size of the global Net Primary Production of the ocean is also similar to that of the land—around 50–60 Pg yr^{-1}. The ocean, however, contains five times as much fixed nitrogen as land does. The implication of ocean and land having similar fluxes but different stock values is that the turnover time on land is five times faster than that in the ocean. A further important difference is that, in the pre-Haber–Bosch era, the land was source of fixed nitrogen, due to river output, while the ocean was a sink. The added industrial and agricultural nitrogen fixation means that land is now also a sink while, in the ocean, the added input from rivers has likely resulted in increased denitrification along the coastal seas and in river mouths. The exact amount of this input is, however, still uncertain. What is known is that, together with increased amounts of phosphorus input, it has led to severe cases of eutrophication (enhanced blooms of phytoplankton). However, it is best to take the numbers in Figure 11.9 as not fixed. As more measurements and new methods and models are applied to close the ocean nitrogen cycle (Jickells et al., 2017), changes of the order of 10% or sometimes more in the fluxes are quite possible.

Just a few large rivers account for most of the nitrogen exported to coastal zones globally (Seitzinger et al., 2010). Compare this, for instance, with the global river runoff

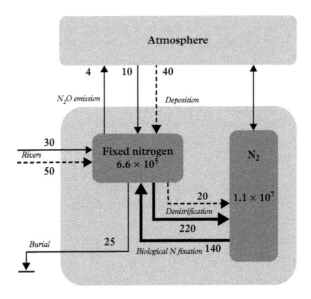

Figure 11.9 *Schematic of the ocean nitrogen cycle. Stocks are in teragrams of nitrogen; fluxes in are in teragrams of nitrogen per year. Nitrogen transformations are given in italics; stocks are given in bold. Dashed lines represent fluxes affected by anthropogenic behaviour. (Values from Gruber & Galloway, 2008)*

shown in Figure 8.9. For example, the twenty-five rivers with the greatest discharge account for approximately half of the exported dissolved inorganic nitrogen (nitrate, nitrite and ammonium), and dissolved organic nitrogen and about a quarter of the exported particulate nitrogen. The dominant source of the nitrogen released into the ocean is nitrate. Generally, the highest nitrogen outflow is by rivers that show considerable anthropogenic input and they thus agree with the areas of high deposition we noted earlier in Europe, the United States, South East Asia, Central Africa and south-east Latin America. However, very large rivers such as the Amazon and Congo dump large amounts of nitrogen in the ocean just by the sheer volume of their discharge. In this case, this occurs mostly in the form of dissolved organic nitrogen.

The input of fixed nitrogen through rivers raises the question of how much of this nitrogen reaches the open ocean. This appears to be a function of the rate of sedimentary denitrification on the shelf and the mean residence time of water on the shelf. The latter, while variable, is subject to constraints of the physics of the ocean flow (see Chapter 6). Remember that we often have currents due to the Coriolis force along the coasts. This suggests that, in cases where the Coriolis force is low, nitrogen may reach the open ocean, while, at higher latitudes, most of the denitrification happens along the shelves. A recent estimate puts the amount of nitrogen reaching the open ocean globally at 75% (Jickells et al., 2017).

Deposition is also a considerable source of added nitrogen. Jickells et al. (2017) estimate that the increased anthropogenic deposition in the ocean is responsible for a 0.4% increase (0.15 Pg C yr^{-1}) in primary production. This is remarkably similar to the enhanced growth in forest (0.18 Pg C yr^{-1}). Both numbers though, carry considerable uncertainty.

11.8 Nitrous oxide from oceans

Figure 11.10 shows an estimate of the global fluxes of nitrous oxide based on work by Freing et al. (2012). The distribution of concentrations and fluxes was obtained using tracer data to calculate transient-time distributions and observation of nitrous oxide from ships. Most of the production of oceanic nitrous oxide is through the nitrification pathway (see Figure 11.1), with denitrification contributing up to 7%–30% of the total. Nitrifying archea are the dominant microbes involved in this. The amount of nitrous oxide produced depends on the availability, or lack, of oxygen in the water column (Voss et al., 2013). Enhanced concentrations of nitrous oxide therefore can be found at oxic–anoxic boundaries in the ocean. In purely anoxic waters, nitrous oxide tends to be further reduced to dinitrogen and is not detectable. In the high Arctic waters, the enhanced solubility of nitrous oxide in cold water shows up as high nitrous oxide concentrations. Figure 11.10 also shows enhanced concentrations in equatorial and coastal upwelling regions with large parts of the open-ocean nitrous oxide concentrations close to zero. Globally, the nitrous oxide emissions from shelf areas are estimated to be less than 0.5%. The data in Figure 11.10 suggest a total emission of 3.1–3.3 Tg N yr^{-1}, close to our estimate of 4 Tg N yr^{-1}, which is based on Figure 11.9.

(a)   (b)

(c)   (d)

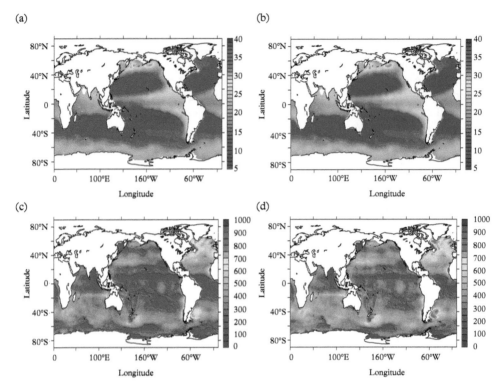

Figure 11.10 *(a, b) Concentration of nitrous oxide in the top 200 metres (in nanomoles per kilogram), as estimated by two slightly different methods. (c, d) Annual flux of nitrous oxide (in micromoles per square metre per year), as estimated by two slightly different methods. (After Freing et al., 2012)*

11.9 The nitrogen cycle in geological times

Denitrification and nitrification processes in the nitrogen cycle are tightly coupled to the availability of oxygen. It is therefore no surprise that the arrival of oxygen on the planet (see Chapter 1) had drastic implications for the evolution of the nitrogen cycle on Earth. To appreciate the evolution of the nitrogen cycle it is again useful to consider the role of the nitrogen isotopes. The stable isotopes of nitrogen (99.64% ^{14}N and 0.36% ^{15}N) are the most widely used tool for reconstructing nitrogen cycling in ancient environments, because they are fractionated by several parts per thousand during many biogeochemical reactions (Stüeken et al., 2016).

Nitrogen fixation in the ocean was generally assumed to cause no fractionation, whereas water column denitrification represents a fractionation of 25‰, which the sedimentary denitrification then has to balance. This had let to estimates for the sedimentary denitrification making up 75% of the total nitrogen conversion. Note that these values depend on the fractionation rates involved. Small changes in either one of those lead immediately lead to changes in the estimated contributions of water column and sediment denitrification. More recent studies suggest that also sedimentary nitrogen loss may involve fractionation.

Now, what can we learn from the geological data (Stüeken et al., 2016)? Before 3.5 Gyr ago, reliable ^{15}N data in rocks is hard to find but, in the Mesoarchean (3.5–2.8 Gyr ago), some more reliable sources are known which suggest a global mean marine sediment value of 1‰ (see Figure 11.11) These values suggest the existence of nitrogen fixation, primarily through nitrogenase based on molybdenum. During this period, the nitrogen cycle must have been anaerobic. In the Archaean oceans, the nitrogen cycle is likely to have been relatively simple (Godfrey & Falkowski, 2009). Dinitrogen would have been biologically reduced to ammonium, which would have become incorporated into organic matter and eventually buried in sediments. Bacterial respiration and grazing would have subsequently released ammonium to the water column to be recycled or buried. At that stage of burial, the tectonic process would have taken over and subducted the nitrogen-containing sediments that could later be brought to the surface. This cycle would have taken millions of years to complete.

This relatively simple cycle changed in the Neoarchean (2.8–2.5 Gyr) with δ^{15}N values becoming more positive, at a global mean of 4.2‰. Nitrates appeared in the ocean. The variability in values found in sedimentary rocks suggest that the nitrogen cycle was still very dynamic in space and time with occasional aerobic periods or places changing quickly to anaerobic places. While the lower values, near 0‰, probably reflect anaerobic nitrogen cycling dominated by nitrogen fixation, the slightly higher high values can be interpreted as evidence of an aerobic nitrogen cycle with nitrification, denitrification and perhaps Anammox as key processes, much as in the modern ocean (Stüeken et al., 2016). The occurrence of slightly enhanced δ^{15}N values coinciding with other indications of temporary oxygenation does provide evidence that biological nitrification and denitrification had evolved by ~2.7 Gyr ago, even though not yet active everywhere on the planet. This would, incidentally, also imply that for, the first time in geological history, nitrogen produced by denitrification would be brought back to atmosphere.

For the next period, the one that includes the Great Oxidation Event at 2.35 Gyr ago, unfortunately little data is available. The data that is available suggest a largely aerobic nitrogen cycle operating as in the Neoarchean, already before the Great Oxidation Event (see also Chapter 1). Other options, such an more anaerobic cycling, cannot be ruled out, but there is a lack of supporting evidence (Stüeken et al., 2016). After the Great Oxidation Event, bulk ^{15}N values in sediment rocks reach sometimes up to 10‰, suggesting enhanced denitrification as a result of high abundance of nitrate.

It was postulated by Canfield (1998) that the Proterozoic ocean was euxinic (i.e. anoxic) and contained sulphide in the form of hydrogen sulphide. In the Mesoproterozoic ocean (1.6–0.8 Gyr ago), this would have led to a ferruginous (see Chapter 1) deep ocean, euxinic continental margins and oxic surface waters. Isotope values of ^{15}N show a peak in bulk values, which can be interpreted as a signal of nitrate availability in nearshore sediments rocks, while a minimum is found in offshore sediment rocks. This suggest also that the ocean was much more nitrate depleted than in previous periods. Nitrogen cycling in the late Neoproterozoic (0.75–0.55 Gyr ago) probably looked very much like that of today, with bulk δ^{15}N values close to today's average.

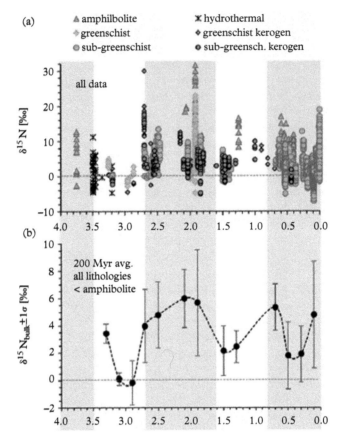

Figure 11.11 *Published δ¹⁵N values, classifed by metamorphic grade (a) and averaged over 0.2 Gyr for only schist and subschist data (b). The bars in (b) represent one standard deviation. Shading marks the periods further discussed in the text. (After Stüeken et al., 2016)*

In the Phanerozoic, $\delta^{15}N$ data suggest a dynamic picture; however, the variability is thought to be caused less by the redox variability, as in previous periods, and more by climatic events. Low values of around 0‰ would originate under warm greenhouse condition with high sea levels and be caused by denitrification from ocean sediments rather than the water column.

11.10 Nitrogen in glacial–interglacial cycles

Ice core measurements present an excellent opportunity to study the Pleistocene nitrogen cycle, as they give us a good record of nitrous oxide during glacial and interglacial cycles. Nitrous oxide results from two central nitrogen cycle processes: nitrification and denitrification.

Values at the last glacial maximum are estimated at 202 ppb, while, at the early Holocene, they had increased up to 270 ppb. For comparison, in September 2017, they were at 330 ppb (see Figure 11.2), the increase since the Holocene due to the use of agriculture and other anthropogenic use. Isotopes suggest a roughly similar contribution in the interglacial rise from terrestrial and marine sources (Figures 11.5 and 11.9), but the precise attribution is still elusive (Gruber & Galloway, 2008). What is clear from the records is that fast fluctuations do occur in line with Dansgaard–Oeschger events, showing a large sensitivity of the nitrogen cycle to climate as in large parts of the Phanerozoic.

The emergence of continental shelves during ice ages due to low sea levels and their flooding during interglacials have been hypothesized to drive changes in sedimentary denitrification. Using data from the South China Sea, Ren et al. (2017) suggest that organic matter decomposition on the shallow shelves during an interglacial promotes high coastal ocean productivity and rapid shelf denitrification. This enhanced denitrification, by consuming fixed nitrogen, causes the shelf water in turn to have excess phosphorus, compared to the standard Redfield ratio (Chapter 9). When this water is transported into the open sea, phytoplankton growth draws down its nutrients, and its excess phosphorus causes nitrogen to become depleted before the phosphorus is. The availability of phosphorus in the absence of nitrogen now, importantly, enhances the nitrogen fixation as a feedback, which is ultimately reflected in a lowering of nitrate $\delta^{15}N$. While these processes can be, to some extent, tested with geochemical data and isotopes, it is good to realize that considerable uncertainties exist in the interpretation of this kind of data.

12

Phosphorus, Sulphur, Iron, Oxygen and Climate

12.1 Phosphorus, sulphur, iron and oxygen

Carbon and nitrogen are arguably the most important elements in the climate system, because their biogeochemical cycles include emissions of gaseous compounds such as carbon dioxide, methane, nitrous oxide and ammonia, and these compounds affect the radiative balance on Earth. However, there are a number of other elements that also play a significant role in the climate system because they affect biomass growth and carbon uptake, processes which ultimately alter the concentrations of carbon dioxide and oxygen in the atmosphere. These elements are phosphorus, sulphur and iron. This chapter covers this important trio of nutrients and their relation with oxygen. Generally, sulphur and phosphorus count as major nutrients, while iron is referred to as a micronutrient.

Before we go into detailed descriptions of the biogeochemistry of these nutrients and their impact on climate at geological timescales, let us first highlight their importance. Phosphorus is used in DNA and RNA and is a key component of energy transfer in organisms through its universal energy carriers adenosine diphosphate (ADP) and adenosine triphosphate (ATP) and is needed to build structural components such as phospholipids and bone material. The lack of available phosphorus is believed to have limited biological productivity in the ocean at geological timescales. On land it limits plant growth, particularly in low-phosphorus soils such as those found in lowland humid tropical forests. Phosphorus is generally not available in an easily accessible form to plants. The process of making phosphorus available for plants is mediated by microbes. The ratio of nitrogen to phosphorus, both on land and in the ocean, is a key indicator of the state of the system and its limitation by lack of phosphorus.

We first encountered iron and sulphur as geological twins in Chapter 1, when we discussed their role as tracers of the oxygenation of the planet. In a reduced world, without oxygen, iron (ferrous) would be ubiquitous in the world's oceans but sulphur (sulphite) almost absent; this is because of their different solubility in water (Chapter 1). In today's oceans, it is the reverse. For iron this had interesting implications for iron which we encountered in Chapter 9 when we discussed the role of dust carrying iron to the ocean

Biogeochemical Cycles and Climate. Han Dolman, Oxford University Press (2019). © Han Dolman.
DOI: 10.1093/oso/9780198779308.001.0001

in the glacial–interglacial cycles. This 'iron hypothesis' was first formulated by John Martin in 1990 (Martin, 1990) and states that enhanced delivery to the Southern Ocean of dust containing iron (see also Chapter 5) during drier glacial periods would lead to increased use of available nutrients (such as nitrogen and phosphorus) and could cause a corresponding drawdown of atmospheric carbon dioxide of up to 50 ppm through enhanced productivity. This idea has led to several proposals in the last few decades to use iron fertilization in the ocean as a way of mitigating climate change. Sulphates and DMS also play important roles in aerosol formation. The use of strato-spheric sulphate aerosols as a means to reflect sunlight is another geoengineering method proposed to combat the effects of extreme climate change. Both this and iron fertilization, as well as the implications of geoengineering, will be further discussed in Chapter 13.

12.2 The phosphorus cycle

The pre-industrial phosphorus cycle is believed to be relatively simple: since no gaseous form exists, weathering of phosphorus-containing rocks provides the input into the world's oceans, where it is used, recycled in the upper layers and a fraction of it ultimately buried. On land, efficient recycling by microbes makes phosphorus available again and again. In places such as Amazonia, however, trans-Atlantic transport of dust may be the only mechanism that provides phosphorus for primary productivity on phosphorus-poor Amazonian soils (Reed et al., 2015).

A schematic of the pre-industrial, geological phosphorus cycle is given in Figure 12.1. Much of the phosphorus is locked in sediments, with soil and ocean stocks holding similar amounts. A much smaller amount is held in terrestrial biomass. Hardly any phos-phorus resides in the atmosphere. Using the definition of residence times from Chapter 1, we indeed find very long mean residence times, on the order of 10,000 years for the ocean, although these times are lower for soil (~200 years). Other estimates come to almost 30,000 years for the ocean, but the size of these estimates depends on the exact size of the stocks and magnitude of the fluxes used, and these inevitably carry a large uncertainty. The long mean residence time in the ocean had led scientists to think that phosphorus only plays a role at geological timescales. That picture has changed over the last decade.

The pre-industrial phosphorus cycle runs in parallel to a cycle of mostly human origin. The phosphorus used in fertilizer is mined from reservoirs of guano deposits (12–17 Tg P yr^{-1}). Part of this, together with other weathering products, ends up in the rivers (in total 13.5–25.0 Tg P yr^{-1}), ultimately enhancing the global riverine input by a factor of 2–3 compared to pre-industrial times. Most of the phosphorus that enters the ocean is in particulate form and it is estimated that about half or two-thirds of the total riverine input is deposited in coastal zones less than 200 m deep. The ranges of the stocks and fluxes shown in Figure 12.1 are considerable, indicating our imprecise and rather uncertain knowledge of the phosphorus cycle at the global scale. Unlike, for instance, water (see Chapter 7), phosphorus is not a renewable resource. It is estimated that, by around

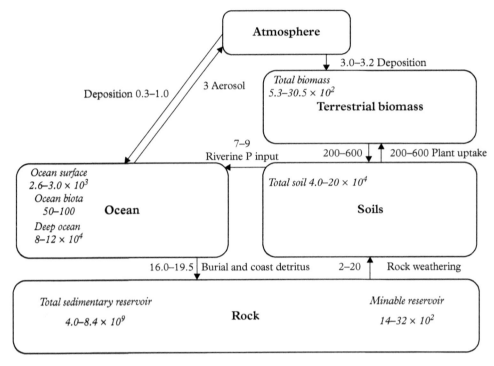

Figure 12.1 *The geological, pre-industrial phosphorus cycle. Fluxes are in teragrams of phosphorus per year; stocks are in teragrams of phosphorus. (Simplified after Peñuelas et al., 2013)*

2050, about 50% of the world's minable phosphorus stock will have been extracted. Then, to be able to sustain current use of phosphorus, ways must be found to recycle human-used phosphorus to almost a full 100%.

12.3 Terrestrial phosphorus

The role of phosphorus in the terrestrial biosphere lies primarily in its impact on photosynthesis and carbon storage. However, there is large variability and phosphorus has been shown to also affect ecosystem structure and biodiversity. The main source of terrestrial phosphorus is bedrock weathering (Figure 12.1), although, in cases where phosphorus is really limited, deposition by dust may also play a role. Since weathering is the main source of phosphorus, over thousands of years of soil development, soils may become depleted in phosphorus. This, albeit a drastic simplification, appears to have happened on tropical humid soils. As nitrogen is generally considered to be fast cycling and available in these forests, phosphorus then becomes the limiting nutrient. One of the important processes in the soil phosphorus cycle is mineralization (Reed et al., 2015), that is, the breakdown of organic phosphorus into mineral forms that can (again) be used by plants. Mineral phosphorus can be used by plants, be absorbed into the soil or be leached. If phosphorus is limited, leaching appears to be small. Absorption of phosphorus is a process that competes with phosphorus utilization by plants and may make it unavailable for decades or centuries. Absorption depends on the type of soil and its chemical composition.

The extra nitrogen input to ecosystems through deposition has significantly changed the classic nitrogen-to-phosphorus ratio of ecosystems as the nitrogen input has increased much faster than the phosphorus input. This has changed the stoichiometric ratio to the disadvantage of phosphorus. While some organisms may adapt to this, not all organisms can adapt easily (Peñuelas et al., 2013), resulting in a changed biodiversity. For instance, increased nitrogen availability on tropical soils would not necessary lead to enhanced primary production, as phosphorus would still be limiting. How this works out precisely is not well enough known to be able to say anything meaningful about potential impacts on climate. Inclusion of phosphorus in climate models suggests that limited phosphorus availability may substantially reduce carbon uptake in the near future. Precisely how the soil processes and microbial activity affect these changes will be key to accurate prediction. Availability of phosphorus or changing nitrogen-to-phosphorus ratios do suggest that current predictions of enhanced carbon uptake because of carbon dioxide fertilization (see Chapter 9) may be over-predictions, and the real uptake may turn out to be substantially smaller when the limitation of nutrients such as nitrogen and phosphorus is taken into account (Peñuelas et al., 2013).

12.4 Phosphorus at geological timescales

We have seen earlier that the nutrients in the modern, pre-industrial ocean by and large obey the Redfield ratio of carbon to nitrogen to phosphorus, 116:16:1. It is, however, very likely that this was not always the case (Planavsky, 2014). For most of Earth's history, the oceans were anoxic and iron rich, as we have seen a number of times throughout this book. Until about half a billion years ago, iron-rich and anoxic waters may have persisted under the oxic top layer. The cycling of nutrients like nitrogen and phosphorus would then have been very different from today, with probably large amounts of bioavailable nitrogen denitrified back to nitrogen gas or through the Anammox process (see also Chapter 11). Such a loss would almost inevitably have led to nitrogen stress for the oceanic biosphere and, to make up for this, an increase in nitrogen fixation in the oxic top layer. This could have changed the ratio of carbon to nitrogen dramatically. Similarly, the phosphorus cycle would be different, with phosphate being trapped in the anoxic layers by the abundant iron. This could also have reduced the overall biological productivity and significantly affected the Redfield ratios.

Figure 12.2 shows a hypothesized scenario that illustrates such changes. It is important to realize that primary productivity was long dominated by cyanobacteria alone and that new plankton groups, such as diatoms, emerged only much more recently. The mineral apatite is the main pathway of phosphorus burial and, as such, is a useful proxy for

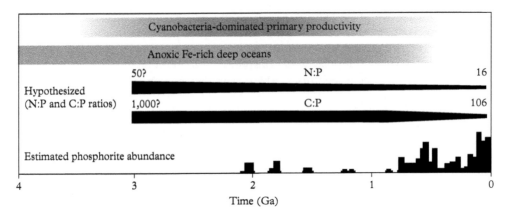

Figure 12.2 *Hypothesized trends in nitrogen-to-phosphorus and carbon-to-phosphorus ratios. (After Planavsky, 2014)*

burial. With high carbon-to-phosphorus ratios (i.e. considerably less phosphorus than carbon), less phosphorus would be buried. In Figure 12.2, the evolution of the nitrogen-to-phosphorus and carbon-to-phosphorus ratios is shown, based on this reasoning. Note the relatively long dominance of the cyanobacteria and anoxic iron-rich deep oceans. On shorter geological timescales, such as in the Cenozoic (~40 Myr ago), phosphorus may also have played an important role, but now it acts very much through erosion and interaction and the carbon–silica cycle. In Chapter 1, we described the geological thermostat, which involved weathering of rocks and had an impact on atmospheric carbon dioxide because of the uptake of one molecule of carbon per one molecule of the mineral wollastonite. For phosphorus, weathering is the key mobilizing agent, and thus changes in weathering rates have the potential to impact the phosphorus cycle and its effect on climate.

Precisely such a hypothesis was formulated in 1989 by Ruddiman and Kutzbach. They proposed that the uplifting of the Himalayas would have led to enhanced weathering releasing more phosphorus and silica into the ocean. The associated drawdown of atmospheric carbon dioxide, because of previous phosphorus limitations, would then have contributed to creating a cooler climate (Ruddiman & Kutzbach, 1989). Part of this impact of the uplift would have been physical: deflection of the jet stream by the sheer height of the plateau, a more intense monsoon and increased rainfall on the slopes of the Himalayas. To determine whether such increased weathering did indeed take place, strontium isotopes—more precisely, the ratio of ^{87}Sr to ^{86}Sr—can be used (see Chapter 3). The ^{87}Sr isotope is produced by radiogenic decay from the metal rubidium. An increase in the strontium isotope ratio represents increased weathering. Over the past 40 million years, the ^{87}Sr/^{86}Sr ratio has steadily increased in the ocean, indicating increased rates of chemical weathering. While this sounds like an attractive hypothesis, it all depends on linking the increase in the strontium ratio to weathering. There are other

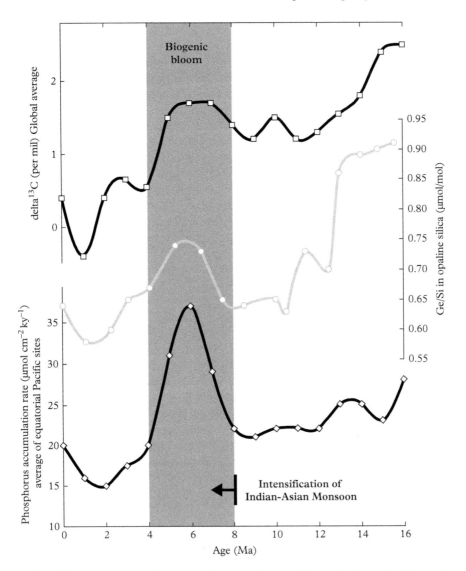

Figure 12.3 *Phosphorus accumulation rate, the germanium-to-silicon ratio in opaline silica, and carbon isotope composition from marine sediments for the past 16 million year. (After Filippelli, 2008)*

processes that could also have produced at least some of the changes and, importantly, the detailed uplift history of the Himalayan Mountains is very not well known.

But let us assume that, after the rise of the Himalayas, chemical weathering did indeed start to increase. What happened next? Increased weathering rates and associated consumption of carbon dioxide when the limitation on phosphorus had gone could imply that the atmosphere would lose virtually all of its carbon dioxide. Yet, there is no indication in the geological record that this, in fact, happened—so, the next question is, why did this not happen? We saw in Chapter 9 that it is possible to determine the burial rate of carbon by measuring carbon isotopes in carbonates and that, geologically, this rate has remained rather constant at 20%, with 80% deposited as carbonate. However, it would appear that, after the rise of the Himalayas, the burial rate of organic carbon did indeed drop, to maybe a value of 10%. Consequently, a net addition of carbon dioxide to the atmosphere was taking place while less carbon was being buried in the ocean sediments. This negative feedback of the carbon cycle would have stopped the decline in carbon dioxide from increased weathering reaching dramatically low levels. Still, the levels dropped and probably significantly enough to have made glaciation of both hemispheres possible. This story of feedbacks, both positive and negative, the use of isotopes as geological tools, and considerable scientific creativity is typical of the way we try to understand paleoclimates and the impact of sudden geological events, such as the rise of the Himalayas, on climate. Obtaining more data from more places may lead to the exclusion of this hypothesis; however, for the time being, it is an attractive one that also shows that not everything on long geological timescales is explained by the geological carbonate–silicate cycle described in Chapter 1.

But where does phosphorus come into play and can we show its importance? Figure 12.3 shows an example of the effects of a modern 'weathering engine', as the Himalayas are sometimes called (Filippelli, 2008). This weathering engine began to be more effective about 8 Myr ago, when the Himalayan uplift became so great that it started to divert the tropical jet stream. This then created the intensified Indian–Asian monsoon system that we mentioned in Chapter 8. Shown in Figure 12.3 are the $\delta^{13}C$ values in sediments, the phosphorus accumulation and the Ge/Si ratio in opal (a matrix of silica and water) in ocean sediments. The latter is also an indication of weathering. Increased weathering would lead to a higher influx of phosphorus, as shown in the accumulation rates. This would stimulate plankton production, because then the main limiting factor would become less stringent and more available—increased carbon burial would then take place. In turn, this could have lowered the concentration of carbon dioxide in the atmosphere, contributing to the onset of the Pleistocene ice ages.

12.5 The iron cycle

The iron cycle is tightly coupled to dust transport (see Chapter 5) and the cycles of phosphorus, nitrogen and sulphur. On land, iron may make up 2%–6% of the soil composition of temperate soils; in the tropical-weathered humid ferral soils, it is much higher. It is virtually never limiting on land under current natural conditions, and its impact on climate is negligible. In contrast, in the ocean, iron is a key limiting resource that shapes the magnitude and dynamics of primary production (Tagliabue et al., 2017). Iron is so scarce in the modern ocean (concentrations at $10^{-12}-10^{-9}$ mol l^{-1}) that it was not until the 1970s that we had reliable values of the iron concentration in the ocean. Now, the improvement in our ability to observe the iron cycle in situ has allowed our understanding of how the ocean iron cycle operates to be refined.

We saw earlier in the discussion of the 'iron hypothesis' in Section 12.1 that iron transport and subsequent fertilization by dust could explain about 50% of the difference between glacial and interglacial carbon dioxide concentrations. The new observations have shown that dust is not the only way that iron can enter the ocean; there are multiple sources of iron in the ocean, including rivers, continental shelves and hydrothermal vents in the mid-ocean ridges. We saw in Chapter 7 that near-the-coast upwelling and transport away from the coast at temperate latitudes are the important pathways of surface water movement. This means that, regionally, sources such as those from continental shelves or mid-ocean ridges are more important than dust. In the Atlantic Ocean, nitrogen is often a limiting nutrient and, when iron is supplied, nitrogen-fixing organisms become more abundant and can make up for the nitrogen deficit. So, iron supply also plays a key role in the nitrogen ocean cycle.

Iron in watery environments forms so-called iron-complexing ligands, complex molecules in which an iron ion is shielded by water molecules that share some of their electrons with the positively charged ion. It is thought that a variety of microbes play an important role in the complexation of iron in the water column. The importance of the process lies in the fact that complexation allows relatively more iron to stay in solution because the water molecules protecting the ion from being scavenged. The deep-water formation of iron-complexing ligands, and the equatorward transport of such ligands from high latitudes, influence the interior distribution of dissolved iron (see Figure 12.4). As these exert a major control on the mean residence time of iron in the ocean, organic iron-complexing ligands are considered to comprise a crucial component of the ocean iron cycle (Tagliabue et al., 2017).

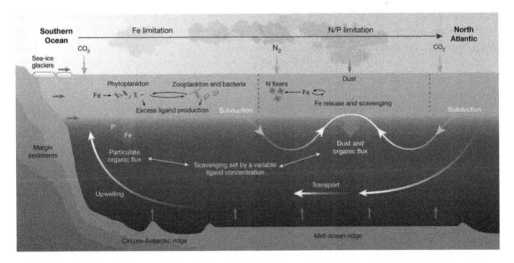

Figure 12.4 *Schematic of the iron cycle in the Atlantic. See text for further explanation. (After Tagliabue et al., 2017)*

Figure 12.4 shows a schematic of the modern-day iron cycle in the ocean. At high latitudes, dust is an important input; at lower latitudes, input from continental shelves and ocean ridges become more important. At high latitudes, the impact of iron is predominantly on the nitrogen cycle. At lower latitudes, such as in the Southern Ocean, upwelling and continental shelves provide much of the input and, in an iron-limited world, the impact is predominantly on the carbon cycle. Because of the low concentration and relatively large fluxes of iron, the mean residence time of iron is on the order of a few tens of years to two hundred years, the range indicating once more our uncertainty about both the size of the inventory and that of the fluxes.

12.6 The sulphur cycle

Sulphur is an important compound in proteins and plays a key role in climate, both in the geological past and in today's climate. We have already seen some of the important roles of DMS and sulphate aerosols in Chapter 5. Sulphur's versatile role in the Earth system's biogeochemistry is made possible by its large variety in redox positions, from S^{2-} in sulphide through to S^{6+} in sulphate. Sulphur in a reduced form is often the nasty-smelling hydrogen sulphide; when oxidized, it is generally sulphate. It is also possible to have a neutral sulphur atom (elemental sulphur, or S^0) in some of the sulphur redox processes. As in the other biogeochemical cycles we have looked at, microbes play a major role in the sulphur cycle, producing a variety of reduced and oxidized products. The ability to oxidize sulphide is present among the Proteobacteria, and it is also found in *Archea*. Many sulphide oxidizers are specially adapted to oxidize sulphide to sulphate through elemental sulphur, such as by building up stores of elemental sulphur in their cells for use when sulphide is lacking in the environment.

At geological timescales, three major pathways exist to transfer sulphur from Earth's mantle into the surface environment and, eventually, the oceans (Canfield & Farquhar, 2012). These are (i) volcanic outgassing of sulphur dioxide and hydrogen sulphide, (ii) the release of hydrogen sulphide from hydrothermal vents in the ocean and (iii) the weathering of sulphide-containing rocks by seawater. The main storage of sulphur is pyrite, a sulphur–iron compound. At geological timescales, it is released again through tectonic processes as sulphur dioxide and hydrogen sulphide. The atmospheric store of sulphur is very small, as particles in the troposphere rain out quickly.

Sulphur is also released when coal and oil are burnt. Since the industrial revolution, humans have thus also significantly perturbed the sulphur cycle, with two major effects. The first was an increase in the amount of acid rain in the previous century. This led to the hypothesis that sulphate emissions reacted with water in the atmosphere to create acid rain, a process which led to acidification of soils and consequent forest dying in Europe and the United States, particularly downwind of industrial areas. When measures were put in place to reduce sulphate emissions, the atmospheric load of sulphate decreased, but the damage to soils (acidification) had already been done and still persists. In some soils, the damage continues to increase, particularly in areas where a reduced buffering capacity of soils is cause for concern. However, it is now thought that the vitality of the affected forests was less damaged than originally thought.

The second effect occurs when sulphur oxidizes to sulphate, as, when it does so, it can form aerosols, which can block the sunlight and can thus have a cooling effect on Earth (see Chapter 5). In fact, scientists believe that the reduction of sulphate aerosols in the 1970s and 1980s made the global warming induced by greenhouse gases become more evident as, before the emissions were reduced, the aerosol contribution 'hid' the warming (Andreae et al., 2005).

Surface or lower tropospheric temperature observations after large volcanic explosions ejecting sulphur-forming aerosols also show a global cooling due to the injection of large amounts of volcanic aerosol in the stratosphere (Robock, 2000). The effects were not just confined to surface cooling, as changes in stratospheric heating via absorption also occurred in parallel, altering the dynamics of the atmosphere. These side effects are of considerable importance for the evaluation of future plans for geoengineering via the injection of stratospheric aerosols, which we will discuss in Chapter 13.

12.7 Iron, sulphur and oxygen in the geological past

Figure 12.5 shows how intricately the evolution of the geological cycle for sulphur is linked to those for oxygen and iron, but it also shows how much we do not yet understand. In particular, note the variation of the $\Delta^{33}S$ fraction: this is the fraction of isotope production expected from mass-dependent fractionation compared to that observed (see Chapter 3). In an oxygen-poor environment, sulphur isotopes, through mass-independent fractionation, would be produced by ultraviolet radiation which, without the substantial ozone shield that we have today, would be able to penetrate deep into the early Earth's atmosphere. The suggestion from these observations is clear: after 2.5 Gyr ago, mass-independent fractionation of sulphur appears to stop (as do the appearances of banded iron formations), a strong indication that, around that time, the atmosphere became substantially oxygenized for the first time in Earth's history (lowest panel, Figure 12.5).

Sulphide can be oxidized by a variety of microbes. However, when the end product, pyrite, is buried in the ocean sediments, a net flux of oxygen to the atmosphere results (Canfield & Farquhar, 2012):

$$2Fe_2O_3 + 16Ca^{2+} + 16HCO_3^- + 8SO_4^{2-} \rightarrow 4FeS_2 + 16CaCO_3 + 8H_2O + 15O_2 \quad (12.1)$$

Equation (12.1) gives the total reaction which has several component reactions, beginning with the production of oxygen and organic carbon through photosynthesis (see eqn (1.1)), oxidation of the organic carbon through sulphate reduction and, finally, the formation of the pyrite, FeS_2. It is thought that the process described by eqn (12.1) was responsible for up to 25% of the oxygen production (carbon burial is thought to be responsible for the remaining 75%; see Chapters 1 and 9) in the Phanerozoic while, in the Proterozoic, this may have been up to 50% (Canfield & Farquhar, 2012).

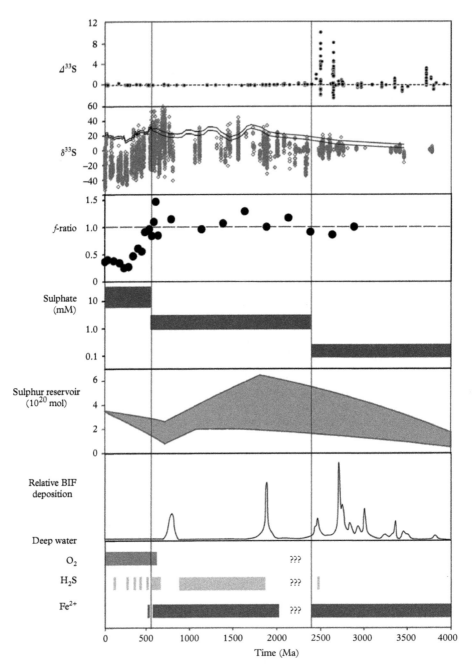

Figure 12.5 *The sulphur cycle in the geological past; BIF, banded iron formations (see Chapter 1). The vertical lines indicate the start of the Cambrian and the Phanerozic at 542 Ma ago, and the Great Oxidation Event around 2400 Ma. The question marks in the bottom indicate lack of data indicating the chemistry in the ocean. (After Canfield & Farquhar, 2012)*

The $\delta^{33}S$ signal in Figure 12.5 indicates that the earliest activity of sulphate reducers appeared at 3.45 Gyr ago. These observations come from analysis of samples of the Dresser Formation in Western Australia (Canfield & Farquhar, 2012). It is worth noting that insight into the biogeochemistry in these times depends on data from a few locations where outcrops of such age are available and, of course, these data may have many interpretations. Nevertheless, it would appear safe to say that the earliest evidence for sulphate-reducing microbes stems from 3.5 Gyr ago. Within a fully oxygenated world, the $\delta^{33}S$ declined again.

How did the sulphur cycle interact with that of iron and oxygen? The iron distribution, particularly in the ocean, was fundamentally different in the Archean, when the oceans and the atmosphere contained little to no oxygen. In that reduced world, iron was abundant in the ocean. It is at this point that the interaction with sulphur comes into play (Raiswell & Canfield, 2012). It appears that Fe^{2+} was in excess as a solute over sulphide in the Archean global ocean, with the main sources of sulphur limited to volcanism and hydrothermal vents. These are minor sources of sulphur today compared to river runoff, which delivers oxidized sources of sulphur to the oceans. In this reduced ocean, sulphate would thus be virtually absent (Raiswell & Canfield, 2012), as we have seen before. Anoxic photosynthesis using light and Fe^{2+} would, under those conditions, be the main source of life-producing biomass.

Now what could have happened next? An increase in oxygen as a result of cyanobacterial photosynthesis during the Great Oxidation Event (2.4–2.5 Gyr ago) would have led to an increase in sulphate concentration due to enhanced oxidative weathering of sulphides on land (the reverse of eqn (12.1)), providing an enhanced source of sulphate to the oceans. We now also assume that, after the Great Oxidation Event, oxygen did not immediately rise to modern levels and was perhaps as low as, or less than, 10% of the present atmospheric level. With some simplification (see Raiswell & Canfield, 2012, for an extensive review), this leads to the conclusion that the increase in sulphate availability as a result of the Great Oxidation Event stimulated rates of sulphate reduction (see also eqn (12.1)) to the point where sulphide production exceeded the delivery of Fe^{2+} to the deep anoxic ocean. This process caused a transition from ferruginous to sulphidic marine conditions. It is very likely that, at the edges of the continents, large anoxic regions would have persisted well into the Proterozoic and that only at a much later stage, 0.5 Gyr ago, would the oceans have become fully oxygenated, containing iron only in its current low amounts.

Associated with the above analysis is the question of what Earth's climate was under those early Archean conditions. Usually, carbon dioxide and methane are invoked to explain how Earth dealt with a sun that produced only about 76%–83% of its current radiation. This we described earlier as the 'faint sun paradox' (Chapter 10). The key to solving that issue is finding the cause for an estimated 50 W m^{-2} of radiative effects! Besides the greenhouse gases, aerosols from DMS and sea spray are sometime invoked, as they may change the size of raindrops and hence structure of the clouds. Other evidence points to ocean temperatures in excess of 50 °C. What is reasonably clear is that the rise of oxygen would have reduced the levels of methane and created the possibility of later glaciations. In this way, the interaction of the biogeochemistry of iron and sulphur and climate is indeed key in our geological history.

12.8 Oxygen in the Pleistocene and the modern world

The story of Earth's oxygenation showed the importance of oxygen in biogeochemical cycling and its link with other cycles, such as those of nitrogen, phosphorus, iron and sulphur. However, oxygen has also played an important diagnostic role in the Pleistocene and still plays such a role in the modern world. It is to these that we now turn our attention. In Chapter 1, we discussed the role of the weathering feedback on global temperatures and suggested that it was hard to prove. Recently, Stolper et al. (2016), using sophisticated analytical equipment, measured the ratio of oxygen to nitrogen in the air trapped in the ice of the Vostok ice cores. They showed that the partial pressure of oxygen has been declining by 7‰ over the past 800,000 years, indicating that the sinks of oxygen outcompete the sources by 2%. This 0.7% decline over 800,000 years was measured against a background of 20% oxygen in the atmosphere. Indeed, the key requirement for measuring these small changes against such a large background of oxygen is a phenomenal precision (Keeling & Manning, 2014; Keeling & Shertz, 1992). The authors were further able to show that this decline would be consistent with changes in sinks, such as the burial of organic material and pyrite, that could be driven by either cooling or increased erosion (Stolper et al. 2016; also see Chapter 1). The fact that the average concentration of carbon dioxide on that 800,000-year timescale was more or less constant (although not on the glacial–interglacial timescale), together with the decline in oxygen, indicates that the silicate-weathering feedback (see Chapter 1), with a timescale of 200,000–500,000 years, might indeed have been operating to keep the concentration of carbon dioxide stable—that is, of course, until humans appeared on the scene, with their ability to use fossil fuel for their energy requirements.

But, at this point in geological history, the role of oxygen is also important. The top panel of Figure 12.6 shows measurements of the oxygen-to-nitrogen ratio at Alert, which is at 82° N, 63° W. These measurements are normalized against an air sample from the mid-eighties, so they show negative values overall. They are expressed in units of molecule per 1 million other molecules (abbreviated as 'per meg'). To be able to get usable numbers, the original values are multiplied by 1000. What we can see is a clear decrease in the value of the ratio since the measurements started in around 1990 (e.g. Keeling & Manning, 2014). This is nothing to worry about in terms of losing the oxygen in the atmosphere that we need to breathe, but the trend indicated by the dashed line is clear. The maximum potential loss of oxygen that can be caused by burning all our fossil fuel reserves is estimated to be is only a few per cent of the total atmospheric content. Importantly, when we plot the concentration of carbon dioxide on the same timescale, we see that is this is steadily increasing, as we have seen a number of times in this book. To be able to burn fossil fuel, we need oxygen, and the declining trend in the oxygen concentration (the concentration of nitrogen remains very much the same) as shown by the changes in the ratio of oxygen to nitrogen must thus in some way be related to the working of the elementary equation in Chapter 1, eqn (1.1). The reality will not exactly follow eqn. (1.1), as the composition of the fossil fuel is not similar to the sugar used in the equation. While the biosphere produces oxygen in the Northern Hemisphere in the summer, we also see the carbon dioxide rising in concert. This causes the periodic

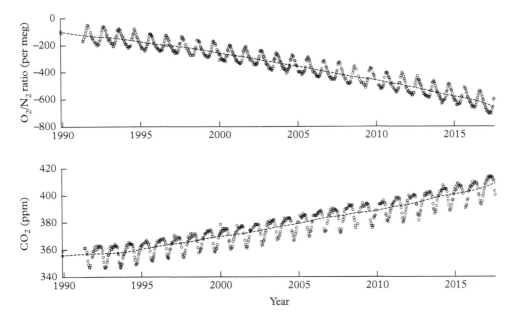

Figure 12.6 *O_2/N_2 ratios and carbon dioxide concentrations, measured at the Scripps research station at Alert, Canada. The dashed trend line was obtained using a stiff spline function. (Data Keeling. R.F. Atmospheric oxygen http://scrippso2.ucsd.edu/osub2sub-data, obtained at 24 May 2018)*

wavelike pattern in Figure 12.6. From eqn (1.1) we also learn that, in photosynthesis and in the reverse process, respiration, on average, one molecule of oxygen is exchanged for one molecule of carbon dioxide. However, the land can also store carbon dioxide in the form of biomass and then lose it via deforestation or changes in land use or deforestation. In the ocean, oxygen has low solubility, so that, unlike the case for carbon dioxide, it cannot really act as a substantial buffer. The long-term uptake of carbon dioxide by the ocean is driven by the carbonate system (see Chapter 9) and involves no oxygen.

This opens the way for an elegant analysis of the relative roles of land and ocean in the carbon cycle (Keeling, et al., 1996). The analysis starts by plotting $\delta(O_2/N_2)$ against carbon dioxide concentrations in the case that changes in both were only due to fossil fuel burning. This is the long downward arrow in Figure 12.7. Since we know that the ocean, minus a small amount of outgassing due to recent ocean warming and stratification, does not exchange oxygen, the remaining arrows (the vector summation in the inset in Figure 12.7) provide us with an estimate of how much the land and ocean have contributed to Earth's sink capacity. In this plot, both land and ocean take up around 30% of the carbon dioxide, an observation which is very much in line what we saw in Chapter 10 (the land uptake as seen by this method is a net uptake, so changes in land use have to be taken into account). So far, this method has proven to be one of the more reliable estimates for the partitioning of the land and ocean sink.

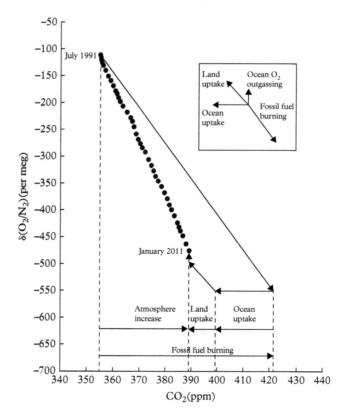

Figure 12.7   *$\delta(O_2/N_2)$ versus carbon dioxide concentrations (both globally and annually averaged) over the period July 1991 through January 2011, computed every six months (solid circles). Data on $\delta(O_2/N_2)$ is from Cape Grim, La Jolla and Alert. (After Keeling and Manning, 2014, following Keeling et al., 1996)*

12.9 Oxygen in the modern ocean

The oxygen that is dissolved in the ocean plays an important role in sustaining life. However, measurements show that oxygen levels in the ocean have been declining since around 1950, due to human activities. Increases in carbon dioxide levels, temperature and, above all, nutrient inputs have changed the distribution of species in the ocean and the composition of seawater. It is estimated that the open ocean has lost 2% of its oxygen over a period of fifty years (Breitburg et al., 2018). This oxygen loss occurs primarily through the expansion of the so-called open-ocean oxygen minimum zones (OMZs), which have expanded by about 4.5 million km². The volume of anoxic water has increased fourfold since 1950. Anoxic waters are waters that show no trace of oxygen; in OMZs, waters have >70 μmol O kg^{-1} at 200 m depth. Ocean warming is held to be responsible for 15% of the decreasing oxygen levels, as it decreases the solubility of oxygen. As ocean

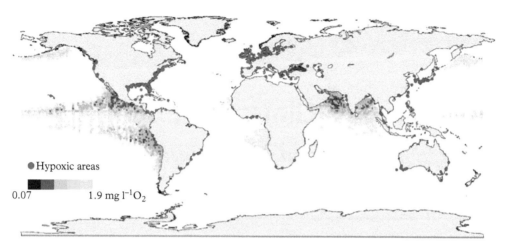

Figure 12.8 *Oxygen levels in the open ocean and coastal waters. The global map indicates coastal sites where anthropogenic nutrients have exacerbated or caused oxygen declines to <2 mg l⁻¹ (dots), as well as ocean oxygen-minimum zones at 300 m depth (shaded regions). (After Breitburg et al., 2018)*

temperatures rise with more than 95% of the greenhouse gas heat going into the ocean, the solubility of oxygen will further decrease. Furthermore, increased temperatures will increase the rates of respiration (Chapter 9) and oxygen consumption, thus blocking transport of oxygen further down into the deeper ocean. So, what causes the remaining 85% of oxygen decline? Increased stratification of the ocean and a weakening overturning circulation (see Chapter 7) are likely responsible for the remaining 85%, with the latter depending on the complex response of wind patterns to global warming (Breitburg et al., 2018).

Figure 12.8 shows the declining oxygen levels in both the open ocean and coastal sites where oxygen levels have declined as a result of nutrient inputs (Breitburg et al., 2018). These estimates are based on so-called repeated hydrographic observations (see Chapter 3). While the coastal-water low oxygen levels correspond with the hot spots of human use of nitrogen (see Chapter 11) and phosphorus, the open-water zones typically are more in the tropical oceans. It is important to realize that changes in OMZs can have regional and global impacts on biogeochemistry through the overturning and upwelling mechanisms of the ocean. As we saw in Chapter 11, nitrous oxide can be produced via both nitrification and denitrification. When oxygen levels are low, nitrous oxide is generally produced; when oxygen becomes absent, nitrogen gas is produced. These are subtle changes, and the sensitivities of key organisms in the biogeochemical cycles of nitrogen, phosphorus and oxygen then become critically important in predicting the future of the oceans and the greenhouse gas content of the atmosphere. In addition, under anoxic conditions, phosphorus and iron may be released from sediments, stimulating biological productivity and thus further oxygen consumption. Such a positive feedback loop may lead to increases in dangerous cyanobacterial blooms, as have occurred, for instance, in the Baltic Sea (Breitburg et al., 2018).

13

The Future of Climate Change

Adaptation, Mitigation, Geoengineering and Decarbonization

13.1 Mitigation, adaptation, geoengineering

To restore a system from a prolonged perturbation, the first action should be to remove that perturbation. This applies equally well to diseases caused by unhealthy lifestyles as it does to climate change caused by burning fossil fuels. To keep warming below acceptable levels, humans need to move away from fossil fuel as their main source and carrier of energy. That being said, such a transition is fraught with problems of equity, acceptance and the (non)-availability of clean technologies. Equity is important in the sense that nations which have growing economies and have emitted little greenhouse gases in their past feel it is unfair that they should share the financial burden of massive reductions of fossil fuel burning with the older economies who have already built their wealth through its use. Humans must accept that changes in lifestyle are needed. This raises issues of readiness and willingness to change one's lifestyle to rely more on clean energy, eat less emission-producing food such as red meat and, for people in growing economies, accept a different future than the one held up as a goal in the old economies. The non-availability of technology to produce clean energy at the required massive global scale, and the current non-availability of techniques to store carbon dioxide or take it out of the air at the required scale, both seriously affect the current assessment of the options for keeping climate from changing further beyond what we now regard as acceptable (1.5–2 °C). Over these difficulties hangs Damocles' sword of time: the longer we wait to make the required changes, the more difficult it will become to do so. For every year of non-decision, the task of limiting climate change to an acceptable level becomes increasingly more difficult, as the Earth system continues to respond to the increased levels of greenhouse gases.

The more-than-a-thousand pages of the IPCC's Working Group I 2013 report provide an excellent and consensus summary of available observations on current and past climate, including our capacity to use Earth system models to predict the future evolution of the climate. In summary, it will get hotter, we will probably experience more

Biogeochemical Cycles and Climate. Han Dolman, Oxford University Press (2019). © Han Dolman.
DOI: 10.1093/oso/9780198779308.001.0001

extremes (drought, floods, heatwaves), sea levels will rise and the oceans will continue to accumulate most of the excess heat and continue to acidify and become more anoxic. As we have seen, Earth has experienced large variability in climates both in the more recent and in the far-away geological past. Over a few tens or hundreds of thousands of years, all greenhouse gases will probably be absorbed back into the ocean, and Earth's climate will be restored to its natural course. While this is likely to happen, the fundamental issue is that human society, as we know it, will no longer be able to function or continue in its present way. Because humans are responsible for this change in climate, it behoves us to think hard about ways we can avoid damaging or catastrophic climate change.

It is useful at this stage to define the terms that are generally used in this discussion more precisely: mitigation, adaptation and geoengineering. We will use the word 'mitigation' to mean reduction of the emissions of greenhouse gases. Continuing with the health metaphor used above, it amounts to taking away the causes of the disease. The word 'adaptation' means one accepts (or has to accept) a continuous perturbation of the system and tries to find ways to adapt to it—trying to live with a disease, say, diabetes, by regularly injecting insulin. In terms of climate, we would think of measures such as raising dikes or making cities greener to battle heatwaves or making them hurricane proof. The symptoms will continue but the economies and livelihoods of the people will adapt to a new climate; this process is foreseen in most climate negotiations. Because it is clear that the current climate has changed, there is already a need for adaptation but, to secure a safe future, mitigation is also required to stop climate reaching a level where the more dangerous impacts can be avoided. If such a crisis were to happen—and, for the moment, we do not exactly define what such a crisis would entail—other measures may be needed. These are generally known by the term 'geoengineering', efforts that deliberately manipulate the planetary environment to counteract anthropogenic climate change. They include schemes such as large-scale injection of sulphate aerosols in the stratosphere, placing solar reflectors at a fixed point between the Sun and Earth, and large-scale catching and dumping of carbon dioxide into the ocean or geological reservoirs. We will describe in this chapter several of the mitigation and adaptation measures and finalize with a discussion of proposed geoengineering schemes.

13.2 Planetary boundaries and adaptation

Planetary boundaries have been suggested as a useful way to look at some of the issues surrounding adaptation (Rockström et al., 2009; Steffen et al., 2015). They define a safe operating space for human society to continue to evolve in a changing environment (Figure 13.1). They also define a level of uncertainty for a particular process that implies the risk of crossing a particular boundary, such as a global temperature threshold. Examples of planetary boundaries are ocean acidification, the nitrogen cycle, plant diversity, ozone loss, freshwater use, aerosol loading and land-cover and land-use change. If the boundaries can be quantified, we can identify where and when we crossed them or below which level we have to remain and adapt to changing conditions. This implies that

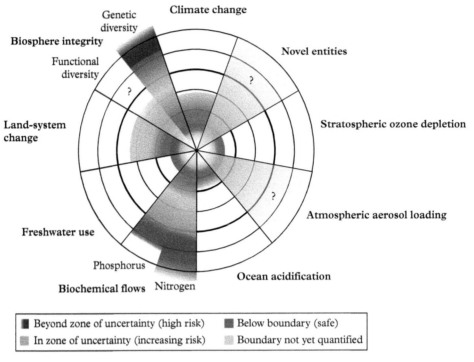

Figure 13.1 *The current status of the nine planetary boundaries. The zone below the first bold circle is the safe operating space (below the boundary), the next represents the zone of uncertainty (increasing risk), and after that is the high-risk zone. Grey wedges with question marks represent processes for which global-level boundaries cannot yet be quantified. (After Steffen et al., 2015)*

we can define a level below which a safe operating space is possible, another level above that where the risks are increasing because we do not know precisely how the system will react or how the drivers may change, and one, above those, which we know is unsafe. They thus define implicitly a level up to where adaptation may be possible.

Some variable needs to be found to represent each Earth system process for which we would like to define a boundary. This will allow monitoring as well as a clear definition of the boundary or zone of uncertainty. For instance, ocean acidification can be represented by the carbonate ion concentration (more specifically, the average global surface–ocean saturation state with respect to aragonite and calcite). This has a critical boundary set at <80% saturation, because, above this value, organisms are not able to form shells and aragonite will start to dissolve (see Chapter 9). Similarly, for climate change, the atmospheric carbon dioxide concentration is a possible variable. An increase in concentration to 350 ppm with a corresponding increase in top-of-the-atmosphere radiative forcing of $+1.0\,W\,m^{-2}$ relative to pre-industrial levels provides the boundary (Steffen et al., 2015). Remember, we have currently just passed the 400 ppm threshold. Although it can be argued that the choice of boundaries is subjective, they provide an important guideline for adaptation and mitigation, and to some extent, research efforts. The boundaries also serve as warnings: if they are crossed, the Earth system is in a state of imbalance.

These characteristics of planetary boundaries make them, in principle, suitable for analysing the interaction of biogeochemical cycles with climate. The carbon cycle climate interaction particularly leads itself to such an analysis, but we could also draw up boundaries for the other cycles discussed in this book. However, the concept is not without criticism. While one can always argue about the level of a particular boundary, such as that for land use and land-use change, or maybe the choice of the representative variable to monitor, other more fundamental criticisms are possible. Let us take, for instance, the choice of the reference period. Should that period be the Holocene, that is, an Earth coming out of a glacial with hardly any human impact, or should it be a pre-industrial world or maybe even the world just after the Second World War? These are not just arbitrary selections of base periods but relate to the fact that the human population not only has changed the environment but is also dependent on it for its food consumption. For instance, it makes virtually no sense to set a nitrogen boundary at a Holocene level of nitrogen fixation, as though the Haber–Bosch process had never been invented. Humans have now become dependent on the use of artificial fertilizer for their food production (de Vries et al., 2013). Humans are both passive inhabitants of the world, suffering from climate change, and, at the same time, active inhabitants changing that very world they inhabit and depend on. Setting planetary boundaries thus turns into a process with moving targets, depending on how much we know and how we want to use and manage the planet.

A second fundamental criticism of the concept is the supposed independency of the individual boundaries. While some may indeed be independent of each other, it is hard to identify a boundary of biodiversity independent of climate change and changes in nitrogen and phosphorus fertilization. In the case of an individual cycle, this becomes even more complicated. Indeed, we have seen many times in this book how complicated the relations between some of the biogeochemical cycles and climate are. For instance, the amount of carbon dioxide in the atmosphere is totally dependent on how much the land and ocean sinks take up. So, subdividing the concentration into its components makes little sense, as one boundary immediately influences the other. The concept that some of the key biogeochemical cycles are moving out or are already beyond the safe operating space is, nevertheless, useful when we describe possible futures of some of the biogeochemical cycles we discussed and how we can use some of the knowledge to mitigate climate change, to keep the planet within a safe operating space.

13.3 Carbon cycle

Let us start with the carbon cycle (see Chapter 8). Figure 13.2 shows the contribution of the main sinks and sources to the change in overall atmospheric inventory since 1870. Note that there is an imbalance at the top, that is, the cumulative sources do not precisely match the cumulative sinks. This is due to the uncertainty involved in the calculation, particularly the sinks and changes due to land use. We start with a value of 288 ppm in

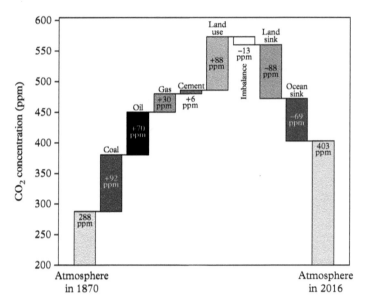

Figure 13.2 *Summary of changes in the global carbon budget since 1870. The budget shown is the cumulative budget for 2016; 1 ppm is equivalent to about 2.2 Pg C. Data: CDIAC/NOAA-ESRL/ GCP/Joos et al 2013. (After GCP (http://www.globalcarbonproject.org/carbonbudget/17/presentation. htm). Data GCP budget 2017 (After Le Quéré et al., 2018))*

1870 and end up with 403 ppm in 2016. Remember that the glacial–interglacial range was 180–280 ppm. The fossil fuel contribution is the largest by 2016, oil and gas being roughly equal to the emissions due to coal. The land-use source, mainly deforestation in the tropics, is considerably less than those other sources, but cumulatively still quite an important factor, due to the accumulated changes in historical land use. In 1950 the cumulative sources from land-use change and fossil fuel were still similar in magnitude; after that, fossil fuel emissions have increased much faster. If the land and ocean sinks had not increased with carbon dioxide concentrations and the resulting climate change, we would have reached about 550 ppm for carbon dioxide concentrations in 2016.

13.3.1 Fossil fuel and allowable emissions

Arguably, Figure 13.3 is the most important graphic to come out of the last IPCC assessment. It shows the near linear relation between cumulative emissions and temperature change relative to the pre-industrial (1861–80) state. The relationship, called the transient climate response to cumulative carbon emissions, relies on the fact that radiative forcing per emitted tonne of carbon dioxide decreases with higher carbon dioxide concentrations (see Chapter 4). This decrease is compensated by the predicted weakening of the land and ocean carbon sinks as they become ever more saturated (see Chapter 9).

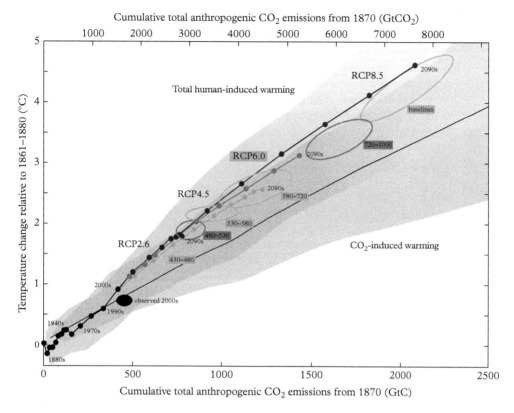

Cumulative total anthropogenic CO$_2$ emissions from 1870 (GtCO$_2$)

Figure 13.3 *The relationship between temperature change, cumulative carbon dioxide emissions and annual greenhouse gas emissions changes by 2050 and 2100 for several emission scenarios (RCP); grey zone indicate uncertainty. Circles indicate CO$_2$ concentration (ppm). (Figure adapted with permission from © 2015 IPCC)*

Overall, this will lead to a larger fraction of emitted carbon dioxide remaining in the atmosphere (Friedlingstein et al., 2014), offsetting the reduced effective radiative warming. In Figure 13.3, the ellipses denote the uncertainty associated with specific emission scenarios. Also shown are the observations of emission and temperature. The linearity between cumulative emissions and temperature has important implications for climate mitigation. As agreed in the Paris Climate Conference of the Parties in 2015, the UN nations committed to avoid climate change above 2 °C, preferably even keeping it below 1.5 °C. In 2014 this would imply that, to stay below 2 °C, emissions around 1000 Pg C would be allowed; for 1.5 °C, this would reduce to about 600 Pg C. All other contributing greenhouse gases are also counted in to this emission potential, so allowable emissions of nitrous oxide and methane will lower the allowable amount of carbon dioxide emissions. Moving several years ahead, the allowable emissions will decrease as the level of cumulative emissions increases. In 2016, for instance, the annual emission from fossil fuel sources and cement production was 9.9 Pg C, and emission due to land-use change was 1.3 Pg C (Le Quéré et al., 2018). The allowable budget thus rapidly dwindles if no mitigation measures are set in place, and keeping global temperature below an agreed threshold becomes increasingly more difficult.

However, by defining what we think is an acceptable level of temperature increase, we also have been able to set, with the help of this graphic, a limit to the amount of fossil fuel that can be burnt, a planetary boundary of some sort. We will come back to this when we discuss the need and ways of mitigation later in this chapter.

13.3.2 Land use

On the same input side of the carbon balance sheet (Figure 13.2), we have land-cover change and land use (management). It is one of the original planetary boundaries and is defined there as land-system change (Steffen et al., 2015). The variable used for this in the original planetary boundaries study is the total areal cover of a specific type of forest. The boundary was set at 54%–75% cover for tropical forest and 85% cover for boreal forest, because of their large impact on climate. Temperate forest was set at 50% cover. It is likely that the effects of land-cover change are thus adequately described. However, land management, the specific way in which humans manage their land, whether it is converted into agricultural crops or left as forest, as it is in large areas in the temperate and boreal regions, also has an impact (Luyssaert et al., 2014). Taking this into account would have the potential to change these boundaries, or impact the current level (see e.g. Erb et al., 2016). In fact, land use has been one of the systems that were considered in the early efforts at climate mitigation, such as the Kyoto Protocol, which was established in 1997. To incorporate this thinking in a carbon cycle framework, it is also necessary to take into account the amount of biomass that must be stored (see Chapter 9, Figure 9.11). That raises the interesting possibility of setting carbon targets for natural vegetation and improving mitigation and carbon sequestration for non-natural land covers.

13.3.3 Land and ocean sinks

In Chapter 9, Figure 9.17, we saw that the land biosphere provides an effective discount on the carbon dioxide emitted by burning fossil fuels, by taking up 29% of those emissions. This land sink, together with the ocean sink, ensure that not all of the emitted carbon dioxide ends up in the atmosphere. The increase in the land sink may come from two positive feedbacks, namely higher temperatures and the increased levels of carbon dioxide in the atmosphere (Figure 9.18). However, increasing temperature affects not only photosynthesis but also respiration and, since it is likely that these processes have different temperature sensitivities, it is not very likely that the land sink will continue to increase, given that, at some point, the trees will not be able to take up more carbon dioxide from the air, while respiration of carbon from the soil will continue to increase. Increasing drought stress may also increase the frequency of forest fires, adding further carbon dioxide to the atmosphere. When systems completely collapse and, for instance, the climate becomes so warm and dry that tropical rainforest can no longer sustain itself

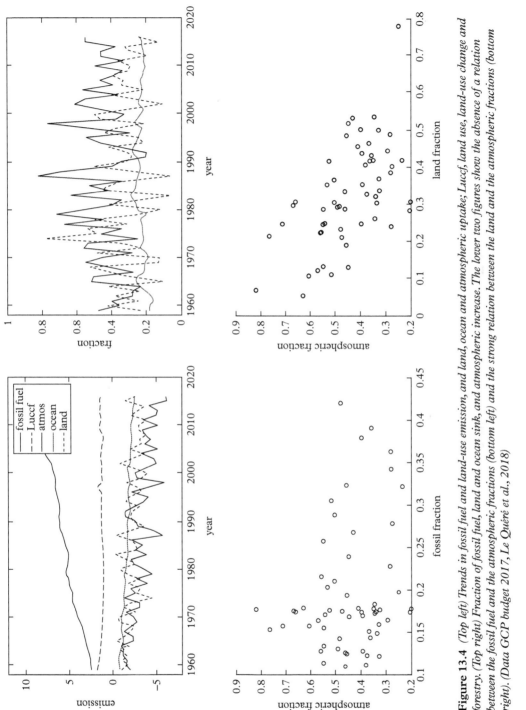

Figure 13.4 *(Top left) Trends in fossil fuel and land-use emission, and land, ocean and atmospheric uptake; Luccf, land-use change and forestry. (Top right) Fraction of fossil fuel, land and ocean sink, and atmospheric increase. The lower two figures show the absence of a relation between the fossil fuel and the atmospheric fractions (bottom left) and the strong relation between the land and the atmospheric fractions (bottom right). (Data GCP budget 2017, Le Quéré et al., 2018)*

(see the moisture recycling discussion in Chapter 7) and is replaced by savanna, this can add substantially to the predicted climate change.

Figure 13.4 shows the carbon budget evolution since 1960, the period for which we have accurate observations of the atmospheric concentration of carbon dioxide (see Chapters 3 and 9). On the top left, we see the trends in sources, fossil fuel and land-use change, and the sinks, ocean, land and atmosphere. As noted before, the land and ocean sinks grow with time, holding the atmospheric increase at around 45% of the fossil fuel emission (top right). This ratio, the atmospheric increase divided by the amount of fossil fuel, is called the atmospheric fraction and, although there is significant interannual variability, it is almost constant over the whole period. While this is, in itself, remarkable, there is no a priori reason why the growth rate of the sinks should keep pace with that of the fossil fuel emissions. It also suggests a useful way to monitor the sink aspect of the carbon cycle. In the lower two panels, the fossil fraction (left) and the land fraction (right) are plotted against the atmospheric fraction. The fossil fuel fraction shows no evidence of a relationship with the atmospheric fraction, but the land fraction does. This suggests that most of the interannual variability we see in the atmospheric carbon dioxide is due to the land. Looking in more detail, we find high growth rates in carbon dioxide during El Niño years.

The ocean sink is more constant, but is also expected to decline because warmer surface water can hold less carbon dioxide, and the decreasing pH shifts the chemical equilibrium towards less uptake. Over the last fifty years, the ocean sink has increased gradually, with the emissions resulting in an almost steady ocean fraction. From a planetary boundary perspective, it would be good to monitor the three fractions: atmospheric, ocean and land. An increase in the atmospheric fraction would show that the fossil fuel emissions are staying more and more in the atmosphere, while a decline in the ocean and land fractions would indicate saturation of those sinks.

13.3.4 Permafrost climate feedback

The permafrost climate feedback deserves special attention within the land sink. In permafrost regions, an estimated 1035 ± 150 Pg C is stored in the frozen soils down to a depth of 3 m (Schuur et al., 2015). In addition, there is the carbon stored in the deep Yedoma regions in Siberia and Alaska; this is estimated to amount to 210 ± 70 Pg or 456 ± 45 Pg, according to two different estimates. An unknown amount is also potentially stored outside the Yedoma regions in other deep layers lying in river deltas (see Figure 10.4). If all of this carbon was released into the atmosphere, this would provide a major feedback on climate, as it is of the same order of magnitude as the allowable remaining fossil fuel emission. This is, however, unlikely to happen. Generally, during the summer season, the warm conditions melt the upper 50 or so cm of the soil. However, more heat means that the thawing depth increases and more organic material is exposed to the air and water. This can produce carbon dioxide, if decomposed aerobically, or methane, if decomposed under anaerobic conditions. Various models estimate that between 34 Pg C and 174 Pg C, with an average of 92 ± 17 Pg C, could be released from this process between now and 2100 (Schuur et al., 2015). This would involve about 10%–15% of the available pool and warm the atmosphere by 0.13–0.27 °C by 2100 and up to 0.42 °C by 2300. This is not a negligible quantity, certainly not in the longer term as the effect accumulates.

13.4 Ocean acidification

Oceanic carbon dioxide uptake has, like the land sink, the apparently benign effect of taking away about a quarter of the fossil fuel emissions, but this comes with a harmful side effect: the conversion of carbon dioxide to carbonate and bicarbonate (see eqns (9.1a, b)) releases protons that acidify the ocean, lowering its pH (see eqn (9.2)). This ocean acidification is often referred to as the 'other carbon dioxide problem'. The pH in the ocean has gone down from 8.2 to 8.1 since pre-industrial times and is expected to go down a further 0.3 or 0.4 units if the concentration of atmospheric carbon dioxide reaches 800 ppm (Doney, et al. 2009). It is worth emphasizing that pH is the logarithm of the H^+ concentration, so the projected drop of 0.3–0.4 units is equivalent to an increase of 150% in the concentration of H^+. Worse, because of the chemical equilibrium between carbon dioxide and bicarbonate in eqn (9.1a), and carbonate and bicarbonate in eqn (9.1b), the concentration of carbonate would go down by 50%. This would have an impact the formation of calcite and aragonite shells, which are responsible for the long-term sequestration of carbon in the ocean sediments. Ocean acidification not only affects the alkalinity of the ocean and calcite formation, but also has impacts on ecosystem health.

Figure 13.5 shows a number of ocean variables, and how they have changed since 1800. The global mean sea level is rising, partly as a result of expansion of ocean waters

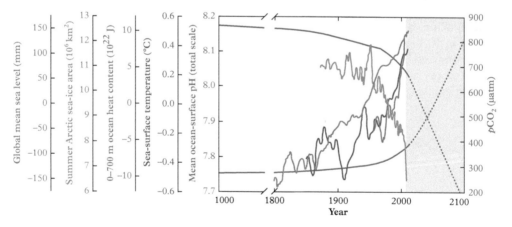

Figure 13.5 *Changes in mean sea level, Arctic summer ice, ocean heat content at 0–700 m, sea-surface temperature and mean ocean surface pH and carbon dioxide concentration in the atmosphere. The shaded area represents the projected changes using the A2 emission scenario of the IPCC Fourth Assessment Report. (After Doney et al., 2012)*

through increased temperature, but also through accumulation of ice loss through glacier melting. Another key variable, the Arctic summer sea ice extent, is declining rapidly, with some expectation that, by sometime in the twenty-first century, it may be gone completely (note that this is summer ice; winter ice cover will not disappear). Most of the excess heat from the greenhouse gas effect is taken up by the ocean, as shown by the increase in heat content. We have already discussed the effects of rising carbon dioxide and pH levels. These are profound changes to the ocean system that cannot be simply reversed or stopped. In fact, if one sets a limit of allowed greenhouse gas emissions that also includes halting ocean acidification, the limits become even more stringent (Steinacher et al., 2014). Taking into account these processes significantly reduces the amount of allowable emissions derived from Figure 13.2 and implies that aiming at a concentration of 500 ppm would be required, which leaves just a small amount of carbon to be emitted under present (2018) conditions.

13.5 Mitigation and negative emissions

What measures are available to mitigate climate change? Next to stopping our fossil fuel use and use renewables, they can be divided into several categories, such as land or biosphere based, geology or, rather, subsurface based and atmosphere based. In Figure 13.6, we identify several of the most important: bioenergy with carbon capture and storage (BECCS); direct air capture of carbon dioxide from ambient air by engineered chemical reactions; enhanced weathering of minerals, where the products are then stored in soils, or buried on land or the deep ocean; afforestation and reforestation; manipulation of carbon uptake by the ocean, either biologically, by fertilizing iron-limited areas, or chemically, by enhancing alkalinity; producing biomass to be locked into products (e.g. houses); and converting biomass to a recalcitrant form called biochar that can be used in soils. Several of these techniques, such as afforestation, have been long known to sequester carbon in the long term; several others, such as BECCS and direct air capture, exist only on the drawing board.

It is worth reiterating the important finding from Figure 13.2 that we have a limited amount of allowable emissions ahead if we want to stay below a 2 °C global temperature

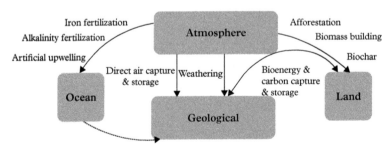

Figure 13.6 *Several measures for mitigation or negative emissions, and the ways they transfer carbon from one reservoir to another.*

increase. Using this amount (around 600 Pg C) as a guide implies that we need to stop emissions almost immediately if we want to avoid overshooting the 2 °C target. Such a drastic reduction is impossible with the current trends in economic thinking and practice that still mostly rely on fossil fuel. Furthermore, the historical growth in fossil fuel emissions shows a continuous near-exponential increase (see Figure 13.4, top left panel) that is unlikely to suddenly break its trend. When trends are broken in this historical record, this has been mostly due to global war (e.g. the Second World War), or large-scale global financial and economic crises (e.g. the 2008 financial crisis).

Some other ways therefore need to be found to curb emission trends. In so-called Integrative Assessment Models (van Vuuren et al., 2013) that are used to develop future scenarios of greenhouse gas emissions based on economic growth, energy projection and land use, this is implemented as a negative emission technology (NET), with which we still will be able to reach the 2 °C target or even the 1.5 °C target. In that case, emissions may continue to grow, but their effect is reduced because we also remove carbon from the air. NETs are based on existing or potential mitigation strategies, but their global use to counter emissions at large scale is very much unproven.

The NETs shown in Figure 13.6 all have different implications in terms of cost, amount of land and energy needed, and amount of water and nutrients needed in the case of the afforestation and biomass-building technologies. Given the task of removing up to 3–5 Pg C yr^{-1} from the atmosphere, these costs quickly become very large. Smith et al. (2015) have calculated the economic and environmental costs of several NETs. Let us take BECCS as a first example. Bioenergy plantations feeding power stations with carbon capture and subsequent storage aimed at 3.3 Pg C yr^{-1} of negative emissions would require a land area of approximately 380–700 Mha in 2100. To compare, such an emission removal would be equivalent to 21% of total human-appropriated Net Primary Production in 2000. Afforestation is estimated to require 320–970 Mha of land. Again, for comparison, the total amount of arable land in 2000 was around 5000 Mha, so BECCS would, if we take an average requirement of about 500 Mha, take up about 10% of the current (!) agricultural land, with a very wide range from 6.4% to 15.8%. Given a growing world population requiring food, this would inevitably lead to unrest in the markets and in societies, unless food production per unit area were to increase dramatically as a result of new agricultural technologies. However, this is not the full story. Plants and trees need water and nutrients. Smith et al. (2015) estimate that, for BECCS alone, the water requirement would be on the order of 10% of the current evapotranspiration from all croplands, or 3% of the human water use of fresh water. These numbers hide the spatial implications: where forest could be grown because land is available, water may not necessarily be available in sufficiently large quantities to sustain forest growth.

If we take a look at some of the other options, such as direct air capture or weathering, it turns out that the energy costs to achieve the required potential for storage are considerable. In the case of direct air capture, taking out 3.3 Pg C yr^{-1} would require about 30% of the projected energy use in 2100. The ocean NETs suffer from similar problems. For instance, adding iron to the ocean will generate a bloom that takes up carbon dioxide, but most of the resulting biomass is then quickly respired again in the upper ocean (see Chapter 9) and little becomes locked in as sediment. The feasibility of NETs living up

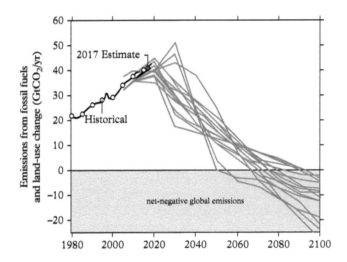

Figure 13.7 *Emission scenarios that stay below a 2 °C limit. The scenarios are based on the Shared Socioeconomic Pathways (Riahi et al., 2017) that are used in the 6th IPCC assessment. Figure from Global Carbon Project, 2017. Le Quéré et al., 2018)*

to their promise presents humanity with a huge uncertainty, because the scenarios used to stay under 2 °C all require a considerable amount of NETs. In Figure 13.7, a set of emission scenarios is shown that allows the planet to stay below the 2 °C limit. These scenarios will be used as input into the sixth IPCC assessment. Next to an almost immediate stop in the growth of emissions, virtually all scenarios require the readiness of substantive NETs around 2040. Turning the trend from increasing emissions to reducing emissions, as laid out in the Paris Agreement, is a huge task and requires massive investment in renewable energy technologies.

13.6 Geoengineering

So, what if we cannot achieve the required changes, if neither NETs nor renewable energy technologies are implemented? To keep Earth from entering into a potentially catastrophic state where, for instance, ice sheets melt and sea levels rise by several metres, some scientists have suggested we should then use geoengineering. Geoengineering, or climate engineering, is defined as the deliberate large-scale manipulation of the planetary environment to counteract anthropogenic climate change. Geoengineering techniques broadly fall into two categories: those dealing with reducing the incoming solar radiation through solar radiation management and those dealing with carbon removal through capture and storage techniques. Before we go on to describe these techniques and their implications in some detail, it is worth discussing the question of what constitutes a climate emergency. Clearly, this is not just a scientific question (Sillmann et al., 2015). The notion of an emergency that can be avoided implies some predictability of that emergency. Say, a sudden loss of the Greenland ice sheet or the decline of tropical forest in Amazonia constitutes what we would call a climate emergency. Such events may be called 'slow tipping points' (see Chapter 1), because the timescale of change is short compared to the timescale of recovery and build-up, that is, climate change progresses at a faster rate than the rate at which the Amazonian forest can be grown back to its original, pre-industrial state. Yet, the predictability is small and we may experience no early warning of the collapse. This makes it inherently difficult to anticipate such an event and set up the large-scale geoengineering system which might prevent it. We will always be too late. Even if we were to apply geoengineering, we might have to apply this in excess to counter hysteresis effects that are common with such events (Sillmann et al., 2015)— it takes considerably more effort to go back to the previous state than just returning to a pre-industrial carbon dioxide level. When we deal with faster processes, such as intensified monsoons or extreme events, these tend to occur on a much more regional basis for which global geoengineering does not necessarily provide the solution. It is also not straightforward to identify a formal cause-and-effect relation between individual extreme events and climate change, although some progress has been made in this area using statistical techniques. In general, declaring a climate emergency to which policymakers would respond with a request for geoengineering requires a political and societal dialogue which has not yet taken place.

Let us now look in more detail at some of the geoengineering techniques. The principle of solar radiation management is easily understood. If we were to reduce the incoming solar radiation by the amount of additional greenhouse forcing for a double carbon dioxide climate, 4 W m^{-2} (see Chapter 4), with an average incoming solar radiation at the surface of 240 W m^{-2}, this would require a 4/240 = 1.7% reduction. This could, in principle, be achieved by flying big mirrors between the Sun and Earth. It is important at this stage to emphasize that this has to be done for a very long time in the future, year in year out, a fact that holds for all geoengineering options. In that respect, it is relevant to look at some longer time perspectives. Figure 13.8 shows trajectories of carbon dioxide and temperature, using intermediate complexity models and emission scenarios for various amounts of carbon dioxide, and carbon dioxide concentrations from ice core data. The key message from this plot is that Earth's system will have to live a very long time with the past, current and future emissions, even if we stop emitting very soon. This effectively means that any effort at solar radiation management will have to continue for at least 10,000 years and probably more, year in year out, unless we manage to take carbon dioxide out of the atmosphere. Any failure to do so will bring about catastrophic warming. This is a very serious commitment to future generations, with large moral and ethical implications.

Apart from direct reflection, the other form of solar management is putting aerosols into the stratosphere. Research on this method of geoengineering was suggested by the winner of the 1995 Nobel Prize for Chemistry, Paul Crutzen, as early as 1990. The idea comes from the effects of large-scale volcanic explosions that have acted in the past to cool the planet. The explosion of Mount Pinatubo in 1991 put enough aerosols in the atmosphere to change the radiative forcing by the required 4 W m^{-2}. We have seen that aerosols may act to cool the planet by reflecting sunlight (see Chapter 5). The geoengineered

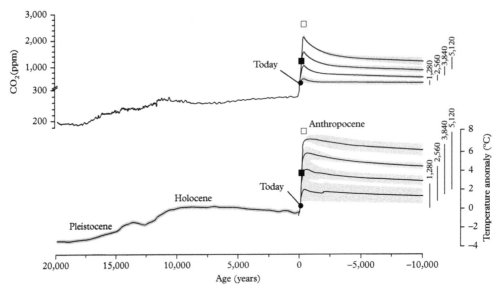

Figure 13.8 *Past and future changes in concentration of atmospheric carbon dioxide and global mean temperature. (Top) Changes in carbon dioxide from ice cores for the past 20,000 years and for four future emission scenarios (1280, 2560, 3840 and 5120 Pg C), from two fully coupled climate–carbon-cycle EMICs. (Bottom) Temperature changes for four future emission scenarios (1280, 2560, 3840 and 5120 Pg C). (After Clark et al., 2016)*

aerosols would have to be brought into the stratosphere by planes and rockets, again for a very long time period. This would have to be repeated, every year, for the next few thousand years. However, these aerosols would also have other effects, such as stratospheric heating, reducing rainfall and, importantly, runoff and intensifying acid rain if we use sulphate aerosols. It is thus difficult to predict their precise regional impact, as we have seen before.

Robock (2008) lists twenty fundamental objections against the use of geoengineering which still hold at the time of writing this chapter (2018). These include, among the ones already mentioned, issues such as governance (i.e. who controls the schemes), the costs, which will need to be shared, and unexpected consequences. It is the latter that is the most worrying and has given rise to some theoretical studies investigating aerosol deployment. Any deliberate experimental effort has met with considerable resistance so far, the main argument being that, by starting to experiment, one would inevitably find oneself on a slippery slope towards real deployment. One further argument is that, if geoengineering were to be set in place, it would then be increasingly difficult to convince nations to mitigate climate change by reducing their emissions. Other ways to change the radiation balance are schemes where sea spray is used to produce enhancement of white clouds over the oceans. The ships doing this would have to be propelled by wave action in the ocean. Exacerbating this issue is climate variability. We have not sufficient understanding of the precise causes of climate variability to accurately predict the effects of these geoengineering schemes. Fundamentally, any tinkering with the radiation budget is dangerous, as our anthropogenic use of fossil fuel has shown.

The only techniques that would help to reduce the amount of carbon dioxide in the atmosphere and keep us from having to stay in the high-carbon-dioxide, high-temperature climate of Figure 13.8 are those that result in direct carbon removal. We have already encountered some of those techniques under mitigation. Unfortunately, these techniques also have some serious issues. Carbon dioxide removal methods do not provide immediate relief and, even if we were able to take out all the excess carbon dioxide immediately, we would still suffer from increased temperatures for some time afterwards due to the inertia of, primarily, the oceans. In principle, the geological stores would be able to store a significant amount of carbon, on the order of several thousands of petagrams of carbon. Carbon stored in the land biomass may be estimated as the potential storage (~900 Pg C) minus the amount that has been lost due to land use and management and has recently been suggested to be at about 50% of the potential, that is, 450 Pg C (Erb et al., 2018). Most of the carbon dioxide removal techniques proposed, such as afforestation or iron fertilization in the ocean, do not have the required capacity to bring down the atmospheric concentration of carbon dioxide. They also lack realistic implementation due to constraints on land, water and so on. Only direct capture of carbon dioxide and subsequent long-term storage would be an option. This has been shown to be feasible so far at one plant in Iceland, where carbon dioxide was mixed with basalt to form calcite (Matter et al., 2016). Surprisingly, the reaction of carbon dioxide with the basalt produced minerals in less than two years. While proven now at small scale, it would require huge investments, and possibly large amounts of energy, to operate this at the global scale.

13.7 Decarbonization of society

The fact that it would be necessary to use NETs to keep the global temperature rise to under 1.5–2.0 °C, and the risks posed by geoengineering techniques, present society with a dilemma that can only be solved by increasing the efforts to change our reliance on fossil fuels and increased efforts at carbon dioxide removal. In short, it requires society to decarbonize fast. This is a daunting but not yet an impossible task.

Let us go back to the discussion on NETs. In the Paris Agreement, individual countries indicated their intention to reduce their greenhouse gas emissions through nationally determined contributions. However, the sum of all of these contributions still leads to an increase in emissions, at least until 2030. This, unfortunately, is in sharp contrast to the pathways required to meet the temperature goals of the Paris Agreement. The temperature target translates into an allowable planetary carbon budget: a 50% chance of limiting warming to 1.5 °C by 2100, and a >66 % probability of meeting the 2 °C target implies that global carbon dioxide emissions need to peak no later than 2020 (two years after writing this book!). Gross emissions need to decline from about 59 Pg CO_2 yr^{-1} in 2020, to 38 Pg CO_2 yr^{-1} by 2030, 22 Pg CO_2 yr^{-1} by 2040 and about 7 Pg CO_2 yr^{-1}

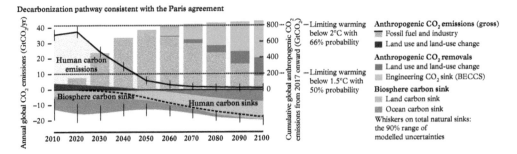

Figure 13.9 *Decarbonization scenario that would cause temperatures to stay below the 2 °C target. See text for further explanation. (After Rockström et al., 2017)*

by 2050 (Rockström et al., 2017). Rockström et al. (2017) have suggested a roadmap for achieving this (note that the values here and in Figure 13.9 are in petagrams of carbon dioxide, which is equivalent to 44/12 = 3.67 Pg C). This roadmap is given schematically in Figure 13.9. They suggest that, initially, the low-hanging fruits should be harvested. This includes stopping subsidies on fossil fuel, stopping coal mining and aggressive funding of renewable energy schemes. This is also the time for cities and states to develop plans to be in the future carbon neutral. The next ten years would be critical to achieving the long-term goal: then the plans need to be implemented, air combustion ways of transport phased out and BECCS developed, along with increased efforts at energy efficiency. In this roadmap, fossil fuel is expected to be phased out completely by 2040. To achieve this would require doubling the use of renewable energy every 5.5 years for the next three decades until 2040. Interestingly, a doubling time of 5.5 years is a historical average doubling time over the period 2005–15, so, in principle, the doubling time is achievable. At some point, also, a carbon tax would probably have to be used to stimulate the transition process. After 2040 the system would need to be revised and strengthened. This roadmap to decarbonization shows that reaching the Paris goals is not unachievable but does require a Herculean effort at the decarbonization of our society.

This roadmap to decarbonization requires massive deployment of NETs that do not yet exist; not as proven technologies at the local scale, and certainly not as technologies that can be implemented at the required global scale and at short notice. The combination of the lack of techniques and the amount of the current nationally determined contributions thus suggests we may still be severely underestimating the issue of climate change. While the Paris Agreement may have raised enthusiasm worldwide at the time it was concluded, reaching the goal is only possible with almost immediate large-scale reduction of emissions through the rapid decarbonizing of our society. This requires changes in the way we define and perceive the interaction between the biogeochemistry of Earth's system and our own economies. In the end, achieving this transition may depend more on the economy and the investment choices made than on the politics of climate negotiations.

14

Reflections on the Anthropocene

We discussed in the previous chapters how humanity is not only affecting the planet's climate but is also now trying to find ways to mitigate climate change. Geoengineering is probably the best example of the search for a solution to such a self-created problem. That is not to say that science should not aim to provide solutions to the set of climate problems we have discussed in this book, but rather that science and, above all, its prac- titioners, the scientists, should be aware of this double role. The definition of planetary boundaries provides another good illustration of this principle. For instance, by defining a pre-industrial level of a variable as a safe operating space, we blissfully ignore the facts that, say fossil fuel use and nitrogen fertilization have provided wealth and well-being to a large number of people on the planet. The choice of a 1.5 or 2 °C level of temperature stabilization is another such case: we agree that a certain level of warming must be accepted to keep our economies going. Similarly, in defining a nitrogen boundary, it makes sense to allow nitrogen fixation for producing food.

All of these human impacts are so fundamental to the functioning of the Earth system that scientists have declared a new geological epoch reflecting these impacts and changes: the Anthropocene (Crutzen, 2002). A committee of geologists has even identified the precise timing and stratigraphic location which is required for such an epoch (Waters et al., 2016). They suggest that it begins when the so-called great acceleration started. During this period, the atmosphere's greenhouse gas concentrations increased strongly, the world started to significantly lose biodiversity and there was an increasing use of plastics and nuclear power. The committee thus put the starting date of the Anthropocene immediately after the Second World War. Key aspects of the Anthropocene are the global and pervasive nature of the change, its omnipresence and the manifold aspects of the change that go beyond just climate change, such as changes in biodiversity and global biogeochemical cycles. As already mentioned, the new two-way interaction between humans and the rest of the natural world is a novel aspect with a planetary scale. It brings a current or imminent fundamental shift in the functioning of our planet as a whole (Malhi, 2017; Steffen, et al., 2007). In virtually all the chapters of this book, we have encountered examples of these shifts.

Since the 1980s, we have had satellites observing Earth; these instruments, combined with new Earth system science approaches, have changed our perspective of the planet. This change in perception is well reflected in the following statement from the classic

Biogeochemical Cycles and Climate. Han Dolman, Oxford University Press (2019). © Han Dolman.
DOI: 10.1093/oso/9780198779308.001.0001

UN Brundtland report, *Our Common Future* (World Commission on Environment and Development, 1987):

> In the middle of the 20th century, we saw our planet from space for the first time. Historians may eventually find that this vision had a greater impact on thought than did the Copernican revolution of the 16th century, which upset the human self-image by revealing that the Earth is not the centre of the universe. From space, we see a small and fragile ball dominated not by human activity and edifice but by a pattern of clouds, oceans, greenery, and soils. Humanity's inability to fit its activities into that pattern is changing planetary systems, fundamentally. Many such changes are accompanied by life-threatening hazards. This new reality, from which there is no escape, must be recognized—and managed.

While the report paved the way for new thinking on sustainable development, it still accepted that Earth's resources were there to be used, albeit managed in a responsible way. Central to the concept of the Anthropocene is the fundamental overall responsibility that humans have to Earth. This responsibility is different from that of previous generations, where it was often more teleological, religious or moral based. What was previously conceived as an environmental crisis has, in the Anthropocene, become a crisis of the planet.

The Anthropocene is different. It is based on the logic of Earth system science: anything we as humans do affects Earth, and Earth's response is by no means passive. From an Earth system science perspective, the key feature is that human domination has led to the emergence of strong feedbacks between humans and the Earth system. These feedbacks occur at a planetary scale, such that actions on energy use, land use, food consumption and trade have consequences for the basic functioning of the planet and can potentially destabilize planetary function (Malhi, 2017). Human societies have always been closely coupled to environmental conditions at local scale, but strong feedback at the planetary scale is the novel and peculiar feature of the Anthropocene. It is this strong human imprint on planetary climate and biogeochemistry that is the true hallmark of the Anthropocene.

The term Anthropocene resonated remarkably well after it was coined in early 2000. It was picked up quickly by the media and found itself being used by sociologists, anthropologists, economists and political scientists. That has made the term rather wide and loosely defined. Nevertheless, regardless of what one thinks of the use of the term, it has served the purpose of making us realize how fundamentally different humans now stand in relation to Earth.

We saw in Chapter 13 what a roadmap for decarbonization could look like. The final goal of that roadmap, global warming below a fixed limit, cannot be achieved without massive new investment, a new international governance system and, above all, a new economic paradigm. Such a new economic paradigm should be based less on the capitalistic aims of exploiting humans and resources for financial profit and more towards conserving and reusing resources, in short, what is known as a circular economy. In a circular economy, the reuse of biogeochemical compounds is critical, and the success of such an economy is by default less based on capital gains than on conserving the Earth system. Such new investments should be profitable but must aim critically at net zero emissions. We should probably apply a firm timetable which these investments would have to follow to arrive at the zero-emission target (Millar et al., 2018). In terms of governance, climate stabilization must be placed on an equal level with economic development, democracy, human rights, and peace (Rockström et al., 2017). Too often in the past, climate change has been placed too low on the political agendas of the international governance systems.

But let us assume that we can achieve most of these reductions; then, how do we know that these emission reductions have been achieved? While the bookkeeping methods in place suggest a reduction, the place where the reductions really matter is the atmosphere. This would require an internationally coordinated observation programme based on in situ and satellite observations of the greenhouse gases and further development of our current tools, such as inversion models, to identify the sources and sinks. Reducing the uncertainties in estimating fossil fuel emissions, importantly, also requires us to reduce the uncertainties in the underlying natural carbon cycle. The response of ocean and land sinks to climate is highly variable, and understanding and monitoring that variability is critical to quantifying any reduction of fossil fuel emissions.

But we have seen that carbon is not the only chemical affecting the biogeochemistry of the planet. The cycles of nitrogen and phosphorus are two other cycles that have been substantially transformed by humans. It is expected that the world will run out of minable phosphorus within the next fifty years. That implies that recycling phosphorus is key to maintaining agricultural production. For nitrogen, the fact that an individual reactive nitrogen molecule can cascade through the environment, causing havoc throughout the cascade, calls for a more integrated approach to minimizing these effects. For agriculture, the concern is mainly improvement in the efficiency of nitrogen use, with respect to both crops and livestock. Here, once more, the approach based on a circular economy—where it is not the selling of the product that takes central place, but its use and subsequent reuse—provides a roadmap to the future. Reduction of emissions through transport and cars is another requirement, as is the recycling of nitrogen out of waste water and manure. One way to approach the economy of this recycling is to include the potential environmental costs of nitrogen use in the overall product. Currently, environmental costs (so-called externalities) are hardly ever taken into account in the cost of the products. As with carbon management, this requires a new circular economy with new governance structures and less emphasis on capital gains and more on the environment. We should stop using Earth as a resource which is almost free of costs.

That brings us finally back to Vladimir Vernadsky, whom we mentioned in Chapter 1 as the originator of the term 'biosphere'. He was the first to realize the essential open nature of Earth's biosphere as a self-controlling system where dead and living matter interact to determine the composition of Earth's surface, atmosphere and hydrosphere. This reads almost as a circular economy *avant la lettre*. Biogeochemistry is what keeps the Earth system going. Realizing the many interactions between these cycles, humans and climate is the basis for creating a sustainable future for the next generations.

References

Chapter 1

Bolin, B., & Rodhe, H. (1973). A note on the concepts of age distribution and transit time in natural reservoirs. *Tellus*, *25*(1), 58–62. http://doi.org/10.3402/tellusa.v25i1.9644

Ciais, P., Sabine, C., Bala, G., Bopp, L., Brovkin, V., Canadell, J., et al. (2013). Carbon and other biogeochemical cycles. In: *Climate Change 2013: The Physical Science Basis. Contribution of Working Group I to the Fifth Assessment Report of the Intergovernmental Panel on Climate Change* [Stocker, T. F., Qin, D., Plattner, G.-K., Tignor, M., Allen, S. K., Boschung, J., et al. (eds.)]. Cambridge University Press, Cambridge, UK, pp. 465–570.

Crutzen, P. J. (2002). Geology of mankind. *Nature*, *415*(6867), 23. http://doi.org/10.1038/415023a

De Paolo, D. J. (2015). Sustainable carbon emissions: the geologic perspective. *MRS Energy & Sustainability*, *2*(2015), 1–16. http://doi.org/10.1557/mre.2015.10.

Kasting, J. (2010). *How To Find a Habitable Planet*. Princeton University Press, Princeton. 267 pp.

Langmuir, C. H., & Broecker, W. (2012). *How To Build a Habitable Planet: The Story of Earth from the Big Bang to Humankind*. Princeton University Press, Princeton. 692 pp.

Lyons, T. W., Reinhard, C. T., & Planavsky, N. J. (2015). The rise of oxygen in Earth's early ocean and atmosphere. *Nature*, *506*(7488), 307–315. http://doi.org/10.1038/nature13068

Vernadsky, V. I. (2012). *The Biosphere*. Copernicus, New York. 199 pp.

Chapter 2

American Meteorological Society. (2019). Climate. In: *Glossary of Meteorology*. American Meteorological Society, Boston, http://glossary.ametsoc.org/wiki/Climate (accessed 11 January 2018).

Augustin, L., Barbante, C., Barnes, P. R. F., Barnola, J.-M., Bigler, M., Castellano, E., et al. (2004). Eight glacial cycles from an Antarctic ice core. *Nature*, *429*(6992), 623–628. http://doi.org/10.1038/nature02599

Etheridge, D. M., Steele, L. P., Langenfelds, R. L., Francey, R. J., J.-M. Barnola, J.-M., & Morgan, V. I. (1998). Historical CO_2 records from the Law Dome DE08, DE08-2, and DSS ice cores. In *Trends: A Compendium of Data on Global Change*. [Carbon Dioxide Information Analysis Center, Oak Ridge National Laboratory] U. S. Department of Energy, Oak Ridge, http://cdiac.ess-dive.lbl.gov/trends/co2/lawdome.html (accessed 2 January 2019).

Galbraith, E. D., & Eggleston, S. (2017). A lower limit to atmospheric CO2 concentrations over the past 800,000 years. *Nature GeoScience*, *10*(4), 295–298. http://doi.org/10.1038/ngeo2914

Goosse, H. (2015). *Climate System Dynamics and Modeling*. Cambridge University Press, Cambridge, UK. 356 pp.

Petit, J. R., Jouzel, J., Raynaud, D., Barkov, N. I., Barnola, J. M., Basile, I., et al. (1999). Climate and atmospheric history of the past 420,000 years from the Vostok ice core, Antarctica. *Nature, 399*(6735), 429–436. http://doi.org/10.1038/20859

Rohling, E. J., Sluijs, A., Dijkstra, H. A., Köhler, P., van de Wal, R. S. W., von der Heydt, A. S., et al. (2012). Making sense of palaeoclimate sensitivity. *Nature, 491*(7426), 683–691. http://doi.org/10.1038/nature11574

van der Werf, G. R., & Dolman, A. J. (2014). Impact of the Atlantic Multidecadal Oscillation (AMO) on deriving anthropogenic warming rates from the instrumental temperature record. *Earth System Dynamics, 5*(2), 375–382. http://doi.org/10.5194/esd-5-375-2014

World Meteorological Organization. (2017). *Frequently Asked Questions (FAQs).* World Meteorological Organization, Geneva, http://www.wmo.int/pages/prog/wcp/ccl/faqs.php (accessed 11 January 2019).

Zachos, J. (2001). Trends, rhythms, and aberrations in global climate 65 Ma to present. *Science, 292*(5517), 686–693. http://doi.org/10.1126/science.1059412

Zachos, J. C., Dickens, G. R., & Zeebe, R. E. (2008). An early Cenozoic perspective on greenhouse warming and carbon-cycle dynamics. *Nature, 451*(7176), 279–283. http://doi.org/10.1038/nature06588

Zeebe, R. E., Ridgwell, A., & Zachos, J. C. (2016). Anthropogenic carbon release rate unprecedented during the past 66 million years. *Nature GeoScience, 9*(4), 325–329. http://doi.org/10.1038/ngeo2681

Chapter 3

Alley, R. B. (2000). *The Two-Mile Machine: Ice Cores, Abrupt Climate Change and Our Future.* Princeton University Press. Princeton. 229 pp.

Baldocchi, D. (2014). Measuring fluxes of trace gases and energy between ecosystems and the atmosphere: the state and future of the eddy covariance method. *Global Change Biology, 20*(12), 3600–3609. http://doi.org/10.1111/gcb.12649

Goosse, H. (2015). *Climate System Dynamics and Modeling.* Cambridge University Press, Cambridge, UK. 356 pp.

Jakob, C. (2014). Going back to basics. *Nature Climate Change, 4*(12), 1042–1045. http://doi.org/10.1038/nclimate2445

Jones, P. D., & Wigley, T. M. L. (2010). Estimation of global temperature trends: what's important and what isn't. *Climatic Change, 100*(1), 59–69. http://doi.org/10.1007/s10584-010-9836-3

Karl, T. R., Arguez, A., Huang, B., & Lawrimore, J. H. (2015). Possible artifacts of data biases in the recent global surface warming hiatus. *Science, 348*(6242), 1469–1472. http://doi.org/10.1126/science.aaa5632

Keeling, C. D. (1960). The concentration and isotopic abundance of carbon dioxide in the atmosphere. *Tellus, 12*(2), 200–203. http://doi.org/10.1111/j.2153-3490.1960.tb01300.x

Riser, S. C., Freeland, H. J., Roemmich, D., Wijffels, S., Troisi, A., Belbeoch, M., et al. (2016). Fifteen years of ocean observations with the global Argo array. *Nature Climate Change, 6*(2), 145–153. http://doi.org/10.1038/nclimate2872

Schouten, S., Hopmans, E. C., & Sinninghe Damsté, J. S. (2013). The organic geochemistry of glycerol dialkyl glycerol tetraether lipids: a review. *Organic Geochemistry, 54*(2003), 19–61. http://doi.org/10.1016/j.orggeochem.2012.09.006

Weart, R. S. (2008). *The Discovery of Global Warming.* Harvard University Press, Cambridge, MA. 228 pp.

Chapter 4

Gaevskaya, G. N., Kondratiev, K. Y., & Yakushevskaya, K. E. (1962). Radiative heat flux divergence and heat regime in the lowest layer of the atmosphere. *Archiv für Meteorologie, Geophysik und Bioklimatologie Serie B, 12*(1), 95–108. http://doi.org/10.1007/BF02317955

Hanel, R. A., Conrath, B. J., Kunde, V. G., Prabhakara, C., Revah, I. Salomonson, V. V., et al. (1972). The Nimbus 4 infrared spectroscopy experiment: 1. Calibrated thermal emission spectra, 1–13. *Journal of Geophysical Research, 77*(15), 2629–2641. http://doi.org/10.1007/BF02317955

Intergovernmental Panel on Climate Change. (2013). Summary for policymakers. In: *Climate Change 2013: The Physical Science Basis. Contribution of Working Group I to the Fifth Assessment Report of the Intergovernmental Panel on Climate Change* [Stocker, T. F., Qin, D., Plattner, G.-K., Tignor, M., Allen, S. K., Boschung, J., et al. (eds.)]. Cambridge University Press, Cambridge, UK, pp. 3–29.

Pierrehumbert, R. T. (2010). *Principles of Planetary Climates*. Cambridge University Press, Cambridge, UK. 652 pp.

Stephens, G., Li, J., Wild, M., Clayson, C. A., Loeb, N., Kato, S., et al. (2012). An update on Earth's energy balance in light of the latest global observations. *Nature Geoscience, 5*(10), 691–696. http://doi.org/10.1038/NGEO1580

Taylor, F. W. (2005). *Elementary Climate Physics*. Oxford University Press. Oxford. 209 pp.

Chapter 5

American Meteorological Society. (2019). Aerosol. In: *Glossary of Meteorology*. American Meteorological Society, Boston, http://glossary.ametsoc.org/wiki/Aerosol (accessed 12 January 2018).

Andreae, M. O. (2009). Correlation between cloud condensation nuclei concentration and aerosol optical thickness in remote and polluted regions. *Atmospheric Chemistry and Physics, 9*(2), 543–556. http://doi.org/10.5194/acp-9-543-2009

Boucher, O. (2015). Atmospheric aerosols. In: *Atmospheric Aerosols*. Dordrecht: Springer Netherlands, pp. 9–24.

Boucher, O., Randall, D., Artaxo, P., Bretherton, C., Feingold, G., Forster, P., et al. (2013). Clouds and aerosols. In: *Climate Change 2013: The Physical Science Basis. Contribution of Working Group I to the Fifth Assessment Report of the Intergovernmental Panel on Climate Change* [Stocker, T. F., Qin, D., Plattner, G.-K., Tignor, M., Allen, S. K., J. Boschung, J., et al. (eds.)]. Cambridge University Press, Cambridge, UK, pp. 571–657.

Charlson, R. J., Langner, J., Rodhe, H., Leovy, C. B., & Warren, S. G. (1991). Perturbation of the northern hemisphere radiative balance by backscattering from anthropogenic sulfate aerosols. *Tellus B: Chemical and Physical Meteorology, 43*(4), 152–163. http://doi.org/10.3402/tellusb.v43i4.15404

Charlson, R. J., Lovelock, J. E., Andreae, M. O., & Warren, S. G. (1987). Oceanic phytoplankton, atmospheric sulfur, cloud albedo and climate. *Nature, 326*(654), 655–661. http://doi.org/10.1038/326655a0

Green, T. A., & Hatton, A. D. (2014). The CLAW hypothesis: a new perspective on the role of biogenic sulphur in the regulation of global climate. *Oceanography and Marine Biology: An Annual Review 52*, 315–336.

Intergovernmental Panel on Climate Change. (2013). Summary for policymakers. In: *Climate Change 2013: The Physical Science Basis. Contribution of Working Group I to the Fifth Assessment Report of the Intergovernmental Panel on Climate Change* [Stocker, T. F., Qin, D., Plattner, G.-K., Tignor, M., Allen, S. K., Boschung, J., et al. (eds.)]. Cambridge University Press, Cambridge, UK, pp. 3–29.

Kinne, S., Schulz, M., Textor, C., Guibert, S., Balkanski, Y., Bauer, S. E., et al. (2006). An AeroCom initial assessment: optical properties in aerosol component modules of global models. *Atmospheric Chemistry and Physics*, 6(7), 1815–1834. http://doi.org/10.5194/acp-6-1815-2006

Kohfeld, K. E., Kohfeld, K. E., Harrison, S. P., & Harrison, S. P. (2001). DIRTMAP: the geological record of dust. *Earth Science Reviews*, 54(1–3), 81–114. http://doi.org/10.1016/S0012-8252(01)00042-3

Kulmala, M., Suni, T., Lehtinen, K. E. J., Dal Maso, M., Boy, M., Reissell, A., et al. (2004). A new feedback mechanism linking forests, aerosols, and climate. *Atmospheric Chemistry and Physics*, 4(2), 557–562. http://doi.org/10.5194/acp-4-557-2004

Luterbacher, J., & Pfister, C. (2015). The year without a summer. *Nature GeoScience*, 8(4), 246–248. http://doi.org/10.1038/ngeo2404

Maher, B. A., Prospero, J. M., Mackie, D., Gaiero, D., Hesse, P. P., & Balkanski, Y. (2010). Global connections between aeolian dust, climate and ocean biogeochemistry at the present day and at the last glacial maximum. *Earth Science Reviews*, 99(1–2), 61–97. http://doi.org/10.1016/j.earscirev.2009.12.001

Robock, A. (2007). Volcanic eruptions and climate, 1–29. *Reviews of Geophysics*, 38(2), 191–219. http://doi.org/10.1029/1998RG000054

Rosenfeld, D., Lohmann, U., Raga, G. B., O'Dowd, C. D., Kulmala, M., Fuzzi, S., et al. (2008). Flood or drought: how do aerosols affect precipitation? *Science*, 321(5894), 1309–1313. http://doi.org/10.526/science.560606

Chapter 6

Ahrens, C. D., & Henson, R. (2016). *Meteorology Today*. Cengage Learning, Boston. 586 pp.

Hays, J. D., Imbrie, J., Imbrie, J., & Shackleton, N. J. (1976). Variations in the Earth's orbit: pacemaker of the ice ages. *Science*, 194(4270), 1121–1132. doi:10.1126/science.194.4270.1121

Holton, J. R. (2004). *Introduction to Dynamical Meteorology: Fourth Edition*. Academic Press, Burlington. 535 pp.

Holton, J. R., & Hakim, G. J. (2012). *An Introduction to Dynamic Meteorology: Fifth Edition*. Elsevier, Amsterdam, 532 pp.

McIlveen, R. (1998). *Fundamentals of Weather and Climate*. Stanley Thornes Publishers Ltd, Cheltenham. 497 pp.

Pauluis, O. M. (2015). The global engine that could: how do hydrological processes affect the atmospheric engine? *Science*, 347(6221), 475–476. http://doi.org/10.1126/science.aaa3681

Stuart Chapin, F., III, Matson, P. A., & Vitousek, P. M. (2011). *Principles of Terrestrial Ecosystem Ecology*. Springer, Berlin. 529 pp.

Wallace, J. M., & Hobbs, P. V. (2006). *Atmospheric Science: An Introductory Survey*. Elsevier, Amsterdam. 483 pp.

Yang, H., Zhao, Y., Liu, Z., Li, Q., He, F., & Zhang, Q. (2015). Heat transport compensation in atmosphere and ocean over the past 22,000 years. *Scientific Reports*, 5(1), 1–19. http://doi.org/10.1038/srep16661

Chapter 7

Apel, J. R. (1988). *Principles of Ocean Physics*. International Geophysics Series, Vol. 38. Academic Press, London. 634 pp.

Balmaseda, M. A., Mogensen, K., & Weaver, A. T. (2012). Evaluation of the ECMWF ocean reanalysis system ORAS4. *Quarterly Journal of the Royal Meteorological Society*, *139*(674), 1132–1161. http://doi.org/10.1002/qj.2063

Good, P., Lowe, J. A., Collins, M., & Moufouma-Okia, W. (2008). An objective tropical Atlantic sea surface temperature gradient index for studies of south Amazon dry-season climate variability and change. *Philosophical Transactions of the Royal Society of London B: Biological Sciences*, *363*(1498), 1761–1766. http://doi.org/10.1098/rstb.2007.0024

Kara, A. B. (2003). Mixed layer depth variability over the global ocean. *Journal of Geophysical Research: Oceans*, *108*(C3), 7138–7115. http://doi.org/10.1029/2000JC000736

Rahmstorf, S. (2006). Thermohaline ocean circulation. In: *Encyclopedia of Quaternary Sciences* [Elias, S. A. (ed.)]. Elsevier, Amsterdam, pp. 737–747.

Stewart, R. H. (2008). *Introduction to Physical Oceanography*. http://www.colorado.edu/oclab/sites/default/files/attached-files/stewart_textbook.pdf (accessed 17 March 2017).

Stommel, H. (1961). Thermohaline convection with two stable regimes of flow. *Tellus*, *13*(2), 224–230. doi:10.1111/j.2153-3490.1961.tb00079.x

Chapter 8

Bentamy, A., Piollé, J. F., Grouazel, A., Danielson, R., Gulev, S., Paul, F., et al. (2017). Review and assessment of latent and sensible heat flux accuracy over the global oceans. *Remote Sensing of Environment*, *201*(2017), 196–218. http://doi.org/10.1016/j.rse.2017.08.016

Boucher, O., Randall, D., Artaxo, P., Bretherton, C., Feingold, G., Forster, P., et al. (2013). Clouds and aerosols. In: *Climate Change 2013: The Physical Science Basis. Contribution of Working Group I to the Fifth Assessment Report of the Intergovernmental Panel on Climate Change* [Stocker, T. F., Qin, D., Plattner, G.-K., Tignor, M., Allen, S. K., Boschung, J., et al. (eds.)]. Cambridge University Press, Cambridge, UK, pp. 571–657.

Clark, P. U., Dyke, A. S., Shakun, J. D., Carlson, A. E., Clark, J., Wohlfarth, B., et al. (2009). The last glacial maximum. *Science*, *325*(5941), 710–714. http://doi.org/10.2307/20544259

Dalin, C., Wada, Y., Kastner, T., & Puma, M. J. (2017). Groundwater depletion embedded in international food trade. *Nature*, *543*(7647), 700–704. http://doi.org/10.1038/nature21403

Dee, D. P., Uppala, S. M., Simmons, A. J., Berrisford, P., Poli, P., Kobayashi, S., et al. (2011). The ERA-Interim reanalysis: configuration and performance of the data assimilation system. *Quarterly Journal of the Royal Meteorological Society*, *137*(656), 553–597.

Gimeno, L. (2014). Atmospheric rivers: a mini-review. *Frontiers in Earth Science*, *2*, 1–6. http://doi.org/10.3389/feart.2014.00002/abstract

Gimeno, L., Dominguez, F., Nieto, R., Trigo, R., Drumond, A., Reason, C. J. C., et al. (2016). Major mechanisms of atmospheric moisture transport and their role in extreme precipitation events. *Annual Review of Environment and Resources*, *41*(1), 117–141. http://doi.org/10.1146/annurev-environ-110615-085558

Martens, B., Miralles, D. G., Lievens, H., van der Schalie, R., de Jeu, R. A. M., Fernández-Prieto, D., et al. (2017). GLEAM v3: satellite-based land evaporation and root-zone soil moisture. *Geoscientific Model Development*, *10*(5), 1903–1925. http://doi.org/10.5194/gmd-10-1903-2017

Milliman, J. D., & Farnsworth, K. L. (2009). *River Discharge to the Coastal Ocean*. Cambridge University Press, Cambridge, UK. 394 pp.

Miralles, D. G., van den Berg, M. J., Gash, J. H. C., Parinussa, R. M, de Jeu, R. A. M. et al. (2014). El Niño–La Niña cycle and recent trends in continental evaporation. *Nature Climate Change*, 4(2), 122–126. http://doi.org/10.1038/nclimate2068

Mueller, B., Hirschi, M., Jimenez, C., Ciais, P., Dirmeyer, P. A., Dolman, A. J., et al. (2013). Benchmark products for land evapotranspiration: LandFlux-EVAL multi-data set synthesis. *Hydrology and Earth System Sciences*, 17(1), 3707–3720. http://doi.org/10.5194/hess-17-3707-2013

Rodell, M., Beaudoing, H. K., L'Ecuyer, T. S., Olson, W. S., Famiglietti, J. S., Houser, P. R., et al. (2015). The observed state of the water cycle in the early twenty-first century. *Journal of Climate*, 28(21), 8289–8318. http://doi.org/10.1175/JCLI-D-14-00555.1

Shiklimanov, I. (2005). World fresh water resources. In: *Water in Crisis* [Gleick, P. H. (ed.)]. Oxford University Press, New York, pp. 13–24.

van der Ent, R. J., Savenije, H. H. G., Schaefli, B., & Steele-Dunne, S. C. (2010). Origin and fate of atmospheric moisture over continents. *Water Resources Research*, 46(9), W09525. http://doi.org/10.1029/2010WR009127

Vaughan, D. G., Comiso, J. C., Allison, I., Carrasco, J., Kaser, G., Kwok, R., et al. (2013). Observations: cryosphere. In: *Climate Change 2013: The Physical Science Basis. Contribution of Working Group I to the Fifth Assessment Report of the Intergovernmental Panel on Climate Change* [Stocker, T. F., Qin, D., Plattner, G.-K., Tignor, M., Allen, S. K., Boschung, J., et al. (eds.)]. Cambridge University Press, Cambridge, UK, pp. 317–382.

Vörösmarty, C. J., & Sahagian, D. (2000). Anthropogenic disturbance of the terrestrial water cycle. *BioScience*, 50(9), 753–765. http://doi.org/10.1641/0006-3568

Yu, L., & Weller, R. A. (2007). Objectively analyzed air-sea heat fluxes for the global ice-free oceans (1981–2005). *Bulletin of the American Meteorological Society*, 88(4), 527–539. http://doi.org/10.1175/BAMS-88-4-527

Zelinka, M. D., Klein, S. A., & Hartmann, D. L. (2012). Computing and partitioning cloud feedbacks using cloud property histograms. Part II: attribution to changes in cloud amount, altitude, and optical depth. *Journal of Climate*, 25(11), 3736–3754. http://doi.org/10.1175/JCLI-D-11-00249.1

Chapter 9

Baldocchi, D. (2014). Measuring fluxes of trace gases and energy between ecosystems and the atmosphere: the state and future of the eddy covariance method. *Global Change Biology*, 20(12), 3600–3609. http://doi.org/10.1111/gcb.12649

Barker, S., & Ridgwell, A. (2012). Ocean acidification. *Nature Education Knowledge*, 3(10), 21.

Beerling, D. J., & Royer, D. L. (2011). Convergent Cenozoic CO2 history. *Nature* 4(7), 418–420. http://doi.org/10.1038/ngeo1186

Carvalhais, N., Carvalhais, N., Forkel, M., Forkel, M., Khomik, M., Khomik, M., et al. (2014). Global covariation of carbon turnover times with climate in terrestrial ecosystems. *Nature*, 514(7521), 213–217. http://doi.org/10.1038/nature13731

Ciais, P., Sabine, C., Bala, G., Bopp, L., Brovkin, V., Canadell, J., et al. (2013). Carbon and other biogeochemical cycles. In: *Climate Change 2013: The Physical Science Basis. Contribution of Working Group I to the Fifth Assessment Report of the Intergovernmental Panel on Climate Change* [Stocker, T. F., Qin, D., Plattner, G.-K., Tignor, M., Allen, S. K., Boschung, J,. et al. (eds.)]. Cambridge University Press, Cambridge, UK, pp. 465–570.

Egleston, E. S., Sabine, C. L., & Morel, F. M. M. (2010). Revelle revisited: buffer factors that quantify the response of ocean chemistry to changes in DIC and alkalinity. *Global Biogeochemical Cycles*, *24*(1), GB1002. http://doi.org/10.1029/2008GB003407

Gutjahr, M., Ridgwell, A., Sexton, P. F., Anagnostou, E., Pearson, P. N., Pälike, H., et al. (2017). Very large release of mostly volcanic carbon during the Palaeocene–Eocene Thermal Maximum. *Nature*, *548*(7669), 573–577. http://doi.org/10.1038/nature23646

Herndl, G. J., & Reinthaler, T. (2013). Microbial control of the dark end of the biological pump. *Nature Publishing Group*, *6*(9), 718–724. http://doi.org/10.1038/ngeo1921

Khatiwala, S., Tanhua, T., Mikaloff Fletcher, S., Gerber, M., Doney, S. C., Graven, H. D., et al. (2013). Global ocean storage of anthropogenic carbon. *Biogeosciences*, *10*(4), 2169–2191. http://doi.org/10.5194/bg-10-2169-2013

Langmuir, C. H., & Broecker, W. (2012). *How To Build a Habitable Planet.* Princeton University Press, Princeton. 736 pp.

Le Quéré, C., Andrew, R. M., Friedlingstein, P., Sitch, S., Pongratz, J., Manning, A. C., et al. (2018). Global carbon budget 2017. *Earth System Science Data*, *10*(1), 405–448. http://doi.org/10.5194/essd-10-405-2018

Le Quéré, C., Moriarty, R., Andrew, R. M., Peters, G. P., Ciais, P., Friedlingstein, P., et al. (2014). Global carbon budget 2014. *Earth System Science Data Discussions*, *7*(2), 521–610. http://doi.org/10.5194/essdd-7-521-2014

Liu, Y. Y., van Dijk, A. I. J. M., de Jeu, R. A. M., Canadell, J. G., McCabe, M. F., Evans, J. P., et al. (2015). Recent reversal in loss of global terrestrial biomass. *Nature Climate Change*, *5*(5), 470–474. http://doi.org/10.1038/nclimate2581

Ruddiman, W. F. (2010). *Plows, Plagues, and Petroleum: How Humans Took Control of Climate.* Princeton University Press, Princeton. 241 pp.

Schulze, E.-D., Wirth, C., & Heimann, M. (2000). Managing forests after Kyoto. *Science*, *289*(5487), 2058–2059. http://doi.org/10.1126/science.289.5487.2058

Sigman, D. M., & Boyle, E. A. (2000). Glacial/interglacial variations in atmospheric carbon dioxide. *Nature*, *407*(6806), 859–869. http://doi.org/10.1038/35038000

Sigman, D. M., Hain, M. P., & Haug, G. H. (2010). The polar ocean and glacial cycles in atmospheric CO_2 concentration. *Nature*, *466*(7), 47–55. http://doi.org/10.1038/nature09149

Sitch, S., Huntingford, C., Gedney, N., Levy, P. E., Lomas, M., Piao, S. L., et al. (2008). Evaluation of the terrestrial carbon cycle, future plant geography and climate-carbon cycle feedbacks using five Dynamic Global Vegetation Models (DGVMs). *Global Change Biology*, *14*(9), 2015–2039. http://doi.org/10.1111/j.1365-2486.2008.01626.x

Van der Meer, D. G., Zeebe, R. E., van Hinsbergen, D. J. J., Sluijs, A., Spakman, W., & Torsvik, T. H. (2014). Plate tectonic controls on atmospheric CO2 levels since the Triassic. *Proceedings of the National Academy of Sciences*, *111*(12), 4380–4385. http://doi.org/10.1073/pnas.1315657111

Wallmann, K., & Aloisi, G. (2012). The global carbon cycle: geological processes. In: *Fundamentals of Geobiology*, Vol. 137. [Knoll, A. H., Canfield, D. E., & Konhauser, K. O. (eds.)]. John Wiley & Sons, Ltd, Chichester, pp. 20–35.

Wanninkhof, R. (2014). Relationship between wind speed and gas exchange over the ocean revisited. *Limnology and Oceanography: Methods*, *12*(6), 351–362. http://doi.org/10.4319/lom.2014.12.351

Wanninkhof, R., Park, G. H., Takahashi, T., Sweeney, C., Feely, R., Nojiri, Y., et al. (2013). Global ocean carbon uptake: magnitude, variability and trends. *Biogeosciences*, *10*(3), 1983–2000. http://doi.org/10.5194/bg-10-1983-2013

Zeebe, R. E. (2012). History of seawater carbonate chemistry, atmospheric CO2, and ocean acidification. *Annual Review of Earth and Planetary Sciences*, *40*(1), 141–165. http://doi.org/10.1146/annurev-earth-042711-105521

Zeebe, R. E., Ridgwell, A., & Zachos, J. C. (2016). Anthropogenic carbon release rate unprecedented during the past 66 million years. *Nature GeoScience*, *9*(4), 325–329. http://doi.org/10.1038/ngeo2681

Chapter 10

Arctic Monitoring and Assessment Programme. (2015). *AMAP Assessment 2015: Methane as an Arctic Climate Forcer*. Arctic Monitoring and Assessment Programme, Oslo. 139 pp.

Bloom, A. A., Palmer, P. I., Fraser, A., Reay, D. S., & Frankenberg, C. (2010). Large-scale controls of methanogenesis inferred from methane and gravity spaceborne data. *Science*, *327*(5963), 322–325. http://doi.org/10.1126/science.1175176

Davidson, E. A., & Janssens, I. A., (2006). Temperature sensitivity of soil carbon decomposition and feedbacks to climate change. *Nature*, *440*(7081), 165–173. http://doi.org/10.1038/nature04514

Dean, J. F., Middelburg, J. J., Röckmann, T., Aerts, R., Blauw, L. G., Egger, M., et al. (2018). Methane feedbacks to the global climate system in a warmer world. *Reviews of Geophysics*, *56*(1), 207–250. http://doi.org/10.1002/2017rg000559

Dlugokencky, E. J., Steele, L. P., Lang, P. M., & Masarie, K. A. (1995). Atmospheric methane at Mauna Loa and Barrow observatories: presentation and analysis of in situ measurements. *Journal of Geophysical Research:Atmospheres*, *100*(D11), 23103–23113. http://doi.org/10.1029/95JD0246

Feulner, G. (2012). The faint young Sun problem. *Reviews of Geophysics*, *50*(2), RG2006. http://doi.org/10.1029/2011RG000375

Holgerson, M. A., & Raymond, P. A. (2016). Large contribution to inland water CO2 and CH4 emissions from very small ponds. *Nature GeoScience*, *9*(3), 222–226. http://doi.org/10.1038/ngeo2654

Hugelius, G., Strauss, J., Zubrzycki, S., & Harden, J. W. (2014). Estimated stocks of circumpolar permafrost carbon with quantified uncertainty ranges and identified data gaps. *Biogeosciences*, *11*(23), 6573–6593. http://doi.org/10.5194/bg-11-6573-2014

Kessler, J. D. (2017). The interaction of climate change and methane hydrates. *Reviews of Geophysics*, *55*(1), 126–168. http://doi.org/10.1002/2016RG000534

Kirschke, S., Bousquet, P., Ciais, P., Saunois, M., Canadell, J. G., Dlugokencky, E. J., et al. (2013). Three decades of global methane sources and sinks. *Nature GeoScience*, *6*(10), 813–823. http://doi.org/10.1038/ngeo1955

Lehmann, J., & Kleber, M. (2015). The contentious nature of soil organic matter. *Nature*, *528*(7580), 60–68. http://doi.org/10.1038/nature16069

Lelieveld, J., Crutzen, P. J., & Dentener, F. J. (1998). Changing concentration, lifetime and climate forcing of atmospheric methane. *Tellus B: Chemical and Physical Meteorology*, *50*(2), 128–150. http://doi.org/10.3402/tellusb.v50i2.16030

Megonigal, J. P., Hines, M. E., & Visscher, P. T. (2014). Anaerobic metabolism: linkages to trace gases and aerobic processes. In: *Biogeochemistry*. Treatise on Geochemistry, Vol. 8. [Holland, H. D., & Turekian, K. K. (eds.)]. Elsevier, Amsterdam, pp. 317–424.

Nisbet, E. G., Dlugokencky, E. J., Manning, M. R., Lowry, D., Fisher, R. E., France, J. L., et al. (2016). Rising atmospheric methane: 2007–2014 growth and isotopic shift. *Global Biogeochemical Cycles*, *30*(9), 1356–1370. http://doi:10.1002/2016gb005406

Nisbet, E. G., & Nisbet, R. E. R. (2008). Methane, oxygen, photosynthesis, rubisco and the regulation of the air through time. *Philosophical Transactions of the Royal Society B: Biological Sciences*, *363*(1504), 2745–2754. http://doi.org/10.1098/rstb.2008.0057

Pangala, S. R., Moore, S., Hornibrook, E. R. C., & Gauci, V. (2012). Trees are major conduits for methane egress from tropical forested wetlands. *New Phytologist, 197*(2), 524–531. http://doi.org/10.1111/nph.12031

Petrescu, A. M. R., van Beek, L. P. H., van Huissteden, J., Prigent, C., Sachs, T., Corradi, C. A. R., et al. (2010). Modeling regional to global CH4 emissions of boreal and arctic wetlands. *Global Biogeochemical Cycles, 24*(4), GB4009. http://doi.org/10.1029/2009GB003610

Quiquet, A., Archibald, A. T., Friend, A. D., Chappellaz, J., Levine, J. G., Stone, E. J., et al. (2015). The relative importance of methane sources and sinks over the Last Interglacial period and into the last glaciation. *Quaternary Science Reviews, 112*(C), 1–16, http://doi:10.1016/j.quascirev.2015.01.004

Raghoebarsing, A. A., Smolders, A. J. P., Schmid, M. C., Rijpstra, W. I. C., Wolters-Arts, M., Derksen, J., et al. (2005). Methanotrophic symbionts provide carbon for photosynthesis in peat bogs. *Nature, 436*(7054), 1153–1156. http://doi:10.1038/nature03802

Saunois, M., Bousquet, P., Poulter, B., Peregon, A., Ciais, P., Canadell, J. G., et al. (2016). The global methane budget 2000–2012. *Earth System Science Data Discussions, 8*(2), 697–751. http://doi:10.5194/essd-8-697-2016

van Huissteden, J., Berrittella, C., Parmentier, F. J. W., Mi, Y., Maximov, T. C., & Dolman, A. J. (2011). Methane emissions from permafrost thaw lakes limited by lake drainage, *Nature Climate Change, 1*(5), 119–123. http://doi.org/10.1038/nclimate1101

Worden, J. R., Bloom, A. A., Pandey, S., Jiang, Z., Worden, H. M., Walker, T. W., et al. (2017). Reduced biomass burning emissions reconcile conflicting estimates of the post-2006 atmospheric methane budget. *Nature Communications, 8*(1), 2227. http://doi.org/10.1038/s41467-017-02246-0

Chapter 11

Anantharaman, K., Brown, C. T., Hug, L. A., Sharon, I., Castelle, C. J., Probst, A. J., et al. (2016). Thousands of microbial genomes shed light on interconnected biogeochemical processes in an aquifer system. *Nature Communications, 7*(2016), 13219. http://doi.org/10.1038/ncomms13219

Butterbach-Bahl, K., Baggs, E. M., Dannenmann, M., Kiese, R., & Zechmeister-Boltenstern, S. (2013). Nitrous oxide emissions from soils: how well do we understand the processes and their controls? *Philosophical Transactions of the Royal Society B: Biological Sciences, 368*(1621), 20130122–20130122. http://doi.org/10.1098/rstb.2013.0122

Butterbach-Bahl, K., & Gundersen, P. (2011). Nitrogen processes in terrestrial ecosystems. In: *The European Nitrogen Assesment* [Sutton, M. A., Howard, C. M., Willem Erisman, J., Billen, G., Bleeker, A., Grennfelt, P., et al.]. Cambridge University Press, Cambridge, UK, pp. 99–125.

Canfield, D. E. (1998). A new model for Proterozoic ocean chemistry. *Nature, 396*(6710), 450–453. http://doi.org/10.1038/24839

Ciais, P., Sabine, C., Bala, G., Bopp, L., Brovkin, V., Canadell, J., et al. (2013). Carbon and other biogeochemical cycles. In: *Climate Change 2013: The Physical Science Basis. Contribution of Working Group I to the Fifth Assessment Report of the Intergovernmental Panel on Climate Change* [Stocker, T. F., Qin, D., Plattner, G.-K., Tignor, M., Allen, S. K., Boschung, J., et al. (eds.)]. Cambridge University Press, Cambridge, UK, pp. 465–570.

Ehtesham, E., & Bengtson, P. (2017). Decoupling of soil carbon and nitrogen turnover partly explains increased net ecosystem production in response to nitrogen fertilization. *Scientific Reports, 7*(2017), 46286. http://doi.org/10.1038/srep46286

Erisman, J. W., Sutton, M. A., Galloway, J., & Klimont, Z. (2008). How a century of ammonia synthesis changed the world. *Nature, 1*(10), 636–639. http://doi.org/10.1038/ngeo325

Freing, A., Wallace, D. W. R., & Bange, H. W. (2012). Global oceanic production of nitrous oxide. *Philosophical Transactions of the Royal Society B: Biological Sciences, 367*(1593), 1245–1255. http://doi.org/10.1098/rstb.2011.0360

Galloway, J. N., Aber, J. D., Erisman, J. W., Seitzinger, S. P., Howarth, R. W., Cowling, E. B., et al. (2003). The nitrogen cascade. *BioScience, 53*(4), 341–356. http://doi.org/10.1641/0006-3568(2003)053[0341:TNC]2.0.CO;2

Godfrey, L. V., & Falkowski, P. G. (2009). The cycling and redox state of nitrogen in the Archaean ocean. *Nature Geoscience, 2*(10), 725–729. http://doi.org/10.1038/ngeo633

Gruber, N., & Galloway, J. N. (2008). An Earth-system perspective of the global nitrogen cycle. *Nature, 451*(7176), 293–296. http://doi.org/10.1038/nature06592

Hedin, L. O., Brookshire, E. N. J., Menge, D. N. L., & Barron, A. R. (2009). The nitrogen paradox in tropical forest ecosystems. *Annual Review of Ecology Evolution and Systematics, 40*(1), 613–635. http://doi.org/10.1146/annurev.ecolsys.37.091305.110246

Hertel, O., Reis, S., Skjoth, C. A., Bleeker, A., Harrison, R., Cape, J. N., et al. (2011). Nitrogen processes in the atmosphere. In: *The European Nitrogen Assessment* [Sutton, M. A., Howard, C. M., Willem Erisman, J., Billen, G., Bleeker, A., Grennfelt, P., et al. (eds.)]. Cambridge University Press, Cambridge, UK, pp. 177–207.

Janssens, I. A., Dieleman, W., Luyssaert, S., Subke, J.-A., Reichstein, M., Ceulemans, R., et al. (2010). Reduction of forest soil respiration in response to nitrogen deposition. *Nature GeoScience, 3*(5), 315–322. http://doi.org/10.1038/ngeo844

Janssens, I. A., & Luyssaert, S. (2009). Carbon cycle: nitrogen's carbon bonus. *Nature Geoscience, 2*(5), 318–319. http://doi.org/10.1038/ngeo505

Jickells, T. D., Buitenhuis, E., Altieri, K., Baker, A. R., Capone, D., Duce, R. A., et al. (2017). A reevaluation of the magnitude and impacts of anthropogenic atmospheric nitrogen inputs on the ocean. *Global Biogeochemical Cycles, 31*(2), 289–305. http://doi.org/10.1002/2016gb005586

Kuypers, M. M. M., Sliekers, A. O., Lavik, G., Schmid, M., Jørgensen, B. B., Kuenen, J. G., et al. (2003). Anaerobic ammonium oxidation by anammox bacteria in the Black Sea. *Nature, 422*(6932), 608–611. http://doi.org/10.1038/nature01472

Lamarque, J. F., Dentener, F., McConnell, J., Ro, C. U., Shaw, M., Vet, R., et al. (2013). Multi-model mean nitrogen and sulfur deposition from the Atmospheric Chemistry and Climate Model Intercomparison Project (ACCMIP): evaluation of historical and projected future changes. *Atmospheric Chemistry and Physics, 13*(16), 7997–8018. http://doi.org/10.5194/acp-13-7997-2013

Ren, H., Sigman, D. M., Martinez-Garcia, A., Anderson, R. F., Chen, M.-T., Ravelo, A. C., et al. (2017). Impact of glacial/interglacial sea level change on the ocean nitrogen cycle. *Proceedings of the National Academy of Sciences, 114*(33), E6759–E6766. http://doi.org/10.1073/pnas.1701315114

Schulte-Uebbing, L., & de Vries, W. (2017). Global-scale impacts of nitrogen deposition on tree carbon sequestration in tropical, temperate, and boreal forests: a meta-analysis. *Global Change Biology, 24*(2), e416–e431. http://doi.org/10.1111/gcb.13862

Seitzinger, S. P., Mayorga, E., Bouwman, A. F., Kroese., C., Beursen, A. H. W., Billen G., et al. (2010). Global river nutrient export: a scenario analysis of past and future trends. *Global Biogeochemical Cycles, 24*(4), GB0A08. http://doi.org/10.1029/2009GB003587

Stüeken, E. E., Kipp, M. A., Koehler, M. C., & Buick, R. (2016). The evolution of Earth's biogeo-chemical nitrogen cycle. *Earth Science Reviews, 160*(2016), 1–20. http://doi.org/10.1016/j.earscirev.2016.07.007

Voss, M., Bange, H. W., Dippner, J. W., Middelburg, J. J., Montoya, J. P., & Ward, B. (2013). The marine nitrogen cycle: recent discoveries, uncertainties and the potential relevance of climate change. *Philosophical Transactions of the Royal Society B: Biological Sciences*, *368*(1621), 20130121–20130121. http://doi.org/10.1098/rstb.2013.0121

Chapter 12

Andreae, M. O., Jones, C. D., & Cox, P. M. (2005). Strong present-day aerosol cooling implies a hot future. *Nature*, *435*(7046), 1187–1190. http://doi.org/10.1038/nature03671

Breitburg, D., Levin, L. A., Oschlies, A., Gregoire, M., Chavez, F. P., Conley, D. J., et al. (2018). Declining oxygen in the global ocean and coastal waters. *Science*, *210*(2), 170–171. http://doi.org/10.1126/science.aam7240

Canfield, D. E., & Farquhar, J. (2012). The global sulfur cycle. In: *Fundamentals of Geobiology*, Vol. 23 [Knoll A. H., Canfield, D. E., & Konhauser, K. D. (eds.)]. John Wiley & Sons, Ltd, Chichester, pp. 49–64 http://doi.org/10.1002/9781118280874.ch5

Filippelli, G. M. (2008). The global phosphorus cycle: past, present, and future. *Elements*, *4*(2), 89–95. http://doi.org/10.2113/GSELEMENTS.4.2.89

Keeling, R. F., & Manning, A. C. (2014). Studies of recent changes in atmospheric O2 content. *Treatise on Geochemistry*, *5*(2014), 385–404. http://doi.org/10.1016/b978-0-08-095975-7.00420-4

Keeling, R. F., Piper, S. C., & Heimann, M. (1996). Global and hemispheric CO$_2$ sinks deduced from changes in atmospheric O$_2$ concentration. *Nature*, *381*(6579), 218–221. http://doi.org/10.1038/381218a0

Keeling, R. F., & Shertz, S. R. (1992). Seasonal and interannual variations in atmospheric oxygen and implications for the global carbon cycle. *Nature*, *358*(6389), 723–727. http://doi.org/10.1038/358723a0

Martin, J. H. (1990). Glacial–interglacial CO2 change: the iron hypothesis. *Paleoceanography*, *5*(1), 1–13. http://doi.org/10.1029/pa005i001p00001

Peñuelas, J., Poulter, B., Sardans, J., Ciais, P., van der Velde, M., Bopp, L., et al. (2013). Human-induced nitrogen–phosphorus imbalances alter natural and managed ecosystems across the globe. *Nature Communications*, *4*(2013), 2934–2934. http://doi.org/10.1038/ncomms3934

Planavsky, N. J. (2014). The elements of marine life. *Nature Geoscience*, *7*(12), 855–856. http://doi.org/10.1038/ngeo2307

Raiswell, R., & Canfield, D. E. (2012). The iron biogeochemical cycle past and present. *Geochemical Perspectives*, *1*(1), 1–220. http://doi.org/10.7185/geochempersp.1.1

Reed, S. C., Yang, X., & Thornton, P. E. (2015). Incorporating phosphorus cycling into global modeling efforts: a worthwhile, tractable endeavor. *New Phytologist*, *208*(2), 324–329. http://doi.org/10.1111/nph.13521

Robock, A. (2000). Volcanic eruptions and climate. *Reviews of Geophysics*, *38*(2), 191–219. http://doi.org/10.1029/1998rg000054

Ruddiman, W. F., & Kutzbach, J. E. (1989). Tectonic forcing of the late Cenozoic climate. *Journal of Geophysical Research: Biogeosciences (2005–2012)*, *94*(D15), 18409–18427. http://doi.org/10.1029/JD094iD15p18409

Stolper, D. A., Bender, M. L., Dreyfus, G. B., Yan, Y., & Higgins, J. A. (2016). A Pleistocene ice core record of atmospheric O2 concentrations. *Science*, *353*(6306), 1427–1430. http://doi.org/10.1126/science.aaf5445

Tagliabue, A., Bowie, A. R., Boyd, P. W., Buck, K. N., Johnson, K. S., & Saito, M. A. (2017). The integral role of iron in ocean biogeochemistry. *Nature*, *543*(7643), 51–59. http://doi.org/10.1038/nature21058

Chapter 13

Clark, P. U., Shakun, J. D., Marcott, S. A., Mix, A. C., Eby, M., Kulp, S., et al. (2016). Consequences of twenty-first-century policy for multi-millennial climate and sea-level change. *Nature Climate Change*, *6*(4), 360–369. http://doi.org/10.1038/nclimate2923

de Vries, W., Kros, J., Kroeze, C., & Seitzinger, S. P. (2013). Assessing planetary and regional nitrogen boundaries related to food security and adverse environmental impacts. *Current Opinion in Environmental Sustainability*, *5*(3–4), 392–402. http://doi.org/10.1016/j.cosust.2013.07.004

Doney, S. C., Fabry, V. J., Feely, R. A., & Kleypas, J. A. (2009). Ocean acidification: the other CO_2 problem. *Annual Review of Marine Science*, *1*, 169–192.

Doney, S. C., Ruckelshaus, M., Emmett Duffy, J., Barry, J. P., Chan, F., English, C. A., et al. (2012). Climate change impacts on marine ecosystems. *Annual Review of Marine Science*, *4*, 11–37. http://doi.org/10.1146/annurev-marine-041911-111611

Erb, K.-H., Kastner, T., Plutzar, C., Bais, A. L. S., Carvalhais, N., Fetzel, T., et al. (2018). Unexpectedly large impact of forest management and grazing on global vegetation biomass. *Nature*, *553*(7686), 73–76. http://doi.org/10.1038/nature25138

Erb, K.-H., Luyssaert, S., Meyfroidt, P., Pongratz, J., Don, A., Kloster, S., et al. (2016). Land management: data availability and process understanding for global change studies. *Global Change Biology*, *23*(2), 512–533. http://doi.org/10.1111/gcb.13443

Friedlingstein, P., Andrew, R. M., Rogelj, J., Peters, G. P., Canadell, J. G., Knutti, R., et al. (2014). Persistent growth of CO_2 emissions and implications for reaching climate targets. *Nature GeoScience*, *7*(1), 709–715. http://doi.org/10.1038/ngeo2248

Le Quéré, C., Andrew, R. M., Friedlingstein, P., Sitch, S., Pongratz, J., Manning, A. C., et al. (2018). Global carbon budget 2017. *Earth System Science Data Discussions*, *10*(2018), 405–448. http://doi.org/10.5194/essd-2017-123

Luyssaert, S., Jammet, M., Stoy, P. C., Estel, S., Pongratz, J., Ceschia, E., et al. (2014). Land management and land-cover change have impacts of similar magnitude on surface temperature. *Nature Climate Change*, *4*(5), 389–393. http://doi.org/10.1038/nclimate2196

Matter, J. M., Stute, M., Snæbjornsdottir, S. O., Oelkers, E. H., Gislason, S. R., Aradottir, E. S., et al. (2016). Rapid carbon mineralization for permanent disposal of anthropogenic carbon dioxide emissions. *Science*, *352*(6291), 1312–1314. http://doi.org/10.1126/science.aad8132

Riahi, K., van Vuuren, D. P., Kriegler, E., Edmonds, J., O'Neill, B. C., Fujimori, S., et al. (2017). The shared socioeconomic pathways and their energy, land use, and greenhouse gas emissions implications: an overview. *Global Environmental Change*, *42*(2017), 153–168. http://doi.org/10.1016/j.gloenvcha.2016.05.009

Robock, A. (2008). 20 reasons why geoengineering may be a bad idea. *Bulletin of the Atomic Scientists*, *64*(2), 14–59. http://doi.org/10.2968/064002006

Rockström, J., Gaffney, O., Rogelj, J., Meinshausen, M., Nakicenovic, N., & Schellnhuber, H. J. (2017). A roadmap for rapid decarbonization. *Science*, *355*(6331), 1269–1271. http://doi.org/10.1126/science.aah3443

Rockström, J., Steffen, W., Noone, K., Persson, Å., Chapin, F., Lambin, E., et al. (2009). A safe operating space for humanity. *Nature*, *461*(7263), 472–475. http://doi.org/10.1038/461472a

Schuur, E. A. G., Schuur, E. A. G., McGuire, A. D., McGuire, A. D., Schädel, C., Grosse, G., et al. (2015). Climate change and the permafrost carbon feedback. *Nature, 520*(7546), 171–179. http://doi.org/10.1038/nature14338

Sillmann, J., Lenton, T. M., Levermann, A., Ott, K., Hulme, M., Benduhn, F., et al. (2015). Climate emergencies do not justify engineering the climate. *Nature Climate Change, 5*(4), 290–292. http://doi.org/10.1038/nclimate2539

Smith, P., Davis, S. J., Creutzig, F., Fuss, S., Minx, J., Gabrielle, B., et al. (2015). Biophysical and economic limits to negative CO2 emissions. *Nature Climate Change, 6*(2015), 42–50. http://doi.org/10.1038/nclimate2870

Steffen, W., Richardson, K., Rockström, J., & Cornell, S. E. (2015). Planetary boundaries: guiding human development on a changing planet. *Science, 347*(6223), 1259855–1259855. http://doi.org/10.1126/science.1259855

Steinacher, M., Joos, F., & Stocker, T. F. (2014). Allowable carbon emissions lowered by multiple climate targets. *Nature, 499*(7457), 197–201. http://doi.org/10.1038/nature12269

van Vuuren, D. P., Deetman, S., van Vliet, J., van den Berg, M., van Ruijven, B. J., & Koelbl, B. (2013). The role of negative CO2 emissions for reaching 2 °CL insights from integrated assessment modelling. *Climatic Change, 118*(1), 15–27. http://doi.org/10.1007/s10584-012-0680-5

Chapter 14

Crutzen, P. J. (2002). Geology of mankind. *Nature, 415*(6867), 23. http://doi.org/10.1038/415023a

Malhi, Y. (2017). The concept of the Anthropocene. *Annual Review of Environment and Resources, 42*, 77–104. http://doi.org/10.1146/annurev-environ-102016-060854

Millar, R. J., Hepburn, C., Beddington, J., & Allen, M. R. (2018). Principles to guide investment towards a stable climate. *Nature Climate Change, 8*(1), 2–4. http://doi.org/10.1038/s41558-017-0042-4

Rockstrom, J., Gaffney, O., Rogelj, J., Meinshausen, M., Nakicenovic, N., & Schellnhuber, H. J. (2017). A roadmap for rapid decarbonization. *Science, 355*(6331), 1269–1271. http://doi.org/10.1126/science.aah3443

Steffen, W., Crutzen, P. J., & McNeill, J. R. (2007). The Anthropocene: are humans now overwhelming the great forces of nature. *AMBIO: A Journal of the Human Environment, 36*(8), 614–621. http://doi.org/10.1579/0044-7447(2007)36[614:taahno]2.0.co;2

Waters, C. N., Zalasiewicz, J., Summerhayes, C., Barnosky, A. D., Poirier, C., Gałuszka, A., et al. (2016). The Anthropocene is functionally and stratigraphically distinct from the Holocene. *Science, 351*(6269), aad2622–aad2622. http://doi.org/10.1126/science.aad2622

World Commission on Environment and Development. (1987). *Report of the World Commission on Environment and Development: Our Common Future.* http://www.un-documents.net/our-common-future.pdf (accessed 27 November 2018).

Index

Tables and figures are indicated by an italic *t* and *f* following the page number.